# Genetics Databases

# BIOLOGICAL TECHNIQUES

## A series of Practical Guides to New Methods in Modern Biology

**Series Editor**
DAVID B SATTELLE

Fluorescent and Luminescent Probes for Biological Activity, Second edition
*WT Mason (Editor)*
Computer Analysis of Electrophysiological Signals
*J Dempster*
Planar Lipid Bilayers
*W Hanke & W-R Schlue*
*In Situ* Hybridization Protocols for the Brain
*W Wisden and BJ Morris (Editors)*
Manual of Techniques in Insect Pathology
*LA Lacey (Editor)*
Non-radioactive Labelling: A Practical Introduction
*AJ Garman*

CLASSIC TITLES IN THE SERIES

Microelectrode Methods for Intracellular Recording and Ionophoresis
*RD Purves*
Immunochemical Methods in Cell and Molecular Biology
*RJ Mayer & JH Walker*

# Contents

## 12  Structural Databases    215
### D. Jones

## 13  PKR – the Protein Kinase Resource    241
### M. Gribskov, P. Bourne & C.M. Smith

## 14  Gene Expression Databases    247
### R. Baldock & D. Davidson

# Contributors

Dr. Richard Baldock
MRC Human Genetics Unit, Western General Hospital, Crewe Road, Edinburgh, EH4 2XU, UK

Dr. Martin Bishop
HGMP Resource Centre, Hinxton, Cambridge, CB10 1SB, UK

Dr. D.J. Blake
University of Oxford, Genetics Unit, Department of Biochemistry, South Parks Road, Oxford, OX1 3QU, UK

Dr. Phil Bourne
San Diego Computer Center, PO Box 85608, San Diego, CA 92186-5608, USA

Dr. Philip Bucher
Institut Suisse de Recherches Experimentales sur le Cancer, Ch. Des Boveresses 155, Epalinges s/Lausanne, CH-1066 Vaud, Switzerland

Dr. Richard Cotton
Mutation Research Centre, St. Vincent's Hospital, Fitzroy 3065, Melbourne, Australia

Dr. Duncan Davidson
MRC Human Genetics Unit, Western General Hospital, Crewe Road, Edinburgh, EH4 2XU, UK

Dr. Jörg T. Epplen
Molecular Human Genetics, Faculty of Medicine MA5, Ruhr University, Universitatesstrasse 150, 44780 Bochum, Germany

Dr. Michael Gribskov
San Diego Supercomputer Center, PO Box 85608, San Diego, CA 92186-5608, USA

Dr. R. Guigó
Information Medica, Institut Municipal d'Investigacio Medica, Dr. Aiguader 80, E08003, Barcelona, Spain

**Dr. Des Higgins**
Department of Biochemistry, University College, Cork, Ireland ·

**Dr. David Jones**
Department of Biological Sciences, University of Warwick, Coventry, CV4 7AL, UK

**Dr. Peter D. Karp**
Senior Scientist, Pangea Systems Inc., 4040 Campbell Avenue, Menlo Park, CA 94025, USA

**Dr. Chris P. Ponting**
Oxford Centre for Molecular Sciences, University of Oxford, South Parks Road, Oxford, OX1 3RH, UK

**Dr. Edúardo J.M. Santos**
Molecular Human Genetics, Faculty of Medicine MA5, Ruhr University, Universitatesstrasse 150, 44780 Bochum, Germany

**Mr. Christopher M. Smith**
San Diego Computer Center, PO Box 85608, San Diego, CA 92186-5608, USA

**Dr. William R. Taylor**
Division of Mathematical Biology, National Institute for Medical Research, The Ridgeway, Mill Hill, London, NW7 1AA, UK

**Dr. Kathleen Triman**
Department of Biology, Franklin and Marshall College, PO Box 3003, Lancaster PA 17604, USA

**Mr. Gary Williams**
HGMP Resource Centre, Hinxton, Cambridge, CB10 1SB, UK

# Preface

We are living in the middle of an information revolution that is changing all our lives. This revolution is more fundamental than the agricultural and industrial revolutions. The ability to sit in our homes and go shopping anywhere in the world, to browse the range of offerings, and to have items delivered to us is becoming commonplace. But this revolution also deals with information about ourselves, how we differ from others in minute detail, and how the cells that constitute our bodies are able to process the information that enables us to live our lives.

This book represents a fragment of the biological information revolution as we struggle to gather the large number of facts that are required as a prerequisite to understanding life. We are not too far away from this understanding for simple life forms, perhaps ten years will suffice. It is fascinating to be living during this period. The authors of the various chapters have attempted to make some of the methodology accessible so that the techniques may be applied by others to enhance our knowledge.

These are indeed exciting times for molecular genetics. Numerous bacterial genomes have been completely sequenced, as well as yeast and nematode. There are advanced sequencing efforts for other organisms including thale cress, fruit fly, mouse and man as well as important pathogens.

We are poised for a major leap in knowledge resulting from the completion of the genome sequences and the development of a number of highly parallel functional data generation methodologies.

We are able to:

- Characterise all the genes in a genome
- Study the effect of removing a gene
- Assign phenotypes to genes
- Understand where gene products are localised in macromolecular assemblies of RNA and protein within cells
- Study all the biological processes and assign biological functions to the gene products.

Most of this book is about the data and analysis tools required for the characterisation of genes and proteins. It is not possible to identify human genes in genomic sequences with any reasonable degree of accuracy using computational methods. A web site, Genesafe http://www.hgmp.mcr.ac.uk/Genesafe/, has been created to help the gene predictors to collaborate on training and testing sets. Genesafe is about making and distributing common datasets for gene finding. It consists of a set of web pages, a mailing list and sets of data in the ftp site.

The last two chapters hint at the major developments in understanding of biological processes and functions that are being made and will expand explosively in the next few years.

Martin Bishop
June 1999

# 1 Introduction

## Martin Bishop

HGMP Resource Centre, Hinxton, Cambridge, UK

## 1.1 INTERNET RESOURCES

Information of all kinds may be accessed on computers around the world that are linked together by the Internet. It may be quicker to retrieve information from the Internet than to make a trip to the library. Many of the same works will be found in both, for example Webster's Dictionary http://work.ucsd.edu:5141/cgi-bin/http_webster. However, computer access is the only method of getting up to date nucleic acid and protein sequence information. Publication of books with compilations of sequences ceased a number of years ago. Journals publish articles that describe the characterization of sequences but may not publish the entire sequence. Instead, they require that the sequence is deposited electronically. Sequences are still distributed on CD-ROM and can be analysed on computers that are not networked. Using this medium, the information is likely to be 3 months or more out of date. Hence, on-line access to sequence and structure information is the norm and is emphasized in this book.

Access to the Internet resources is best achieved via a graphical user interface. The World Wide Web (WWW) is geared to this and a browser such as Netscape or Microsoft Explorer is required plus a network service provider and suitable hardware. Some software will require X-Windows that is provided on Unix workstations or as emulations on Windows and Macintosh computers. Even the need for X is eliminated by a system known as Virtual Network Computing (VNC) that is run on the applications server with a client on your own machine. Most universities provide suitable facilities for their members, including undergraduates.

The subsequent chapters will describe what is available in detail. In order to find the information that you require there are three main approaches:

(1) You know the WWW address of the resource you require e.g. National Center for Biotechnology Information http://www/ncbi.nlm.nih.gov/
(2) You know a site which compiles information relating to the subject

*Genetics Databases*
ISBN 0-12-101625-0

area of interest with links to other sites, e.g. Weizmann Institute of Science Genome and Bioinformatics http://bioinformatics. weizmann.ac.il/

(3) You use a Web search engine and enter words that you hope will lead you to the relevant information somewhere on the Web e.g. 'protein AND database' query on Excite leads to Proteome, Inc. http://quest7.proteome.com/

## 1.2   ORGANISMS AND PROTEINS

The living world is very diverse but it has become apparent that the similarities between organisms at the molecular level are as striking as the differences at the morphological level. Except for prion particles and RNA viruses, all organisms contain DNA that encodes information which is transcribed into RNA. RNA serves a number of diverse functions but much of it is translated into proteins. This book is concerned with the techniques required to study the biological space of organisms with respect to information about all their constituent DNA, RNA and protein molecules. To document these molecules is the task of the nucleic acid and protein databases. The set of organisms studied is presently restricted by the capacity and cost of current sequencing technology. We may expect this set to be expanded in the future as automation and miniaturization of sequencing is achieved. The objective is far more than a simple catalogue. It is to document the function of each molecule and its interactions with other molecules and to integrate this knowledge into a model for the development and function of the organism. The complexity of this task will clearly vary with the complexity of the organism. The beauty of these studies is that lessons learned from simple organisms can be applied to the understanding of more complex organisms. For example, a fundamental question in biology is 'how does an animal develop from a single cell into a functioning adult?' Many aspects of development and function are similar in animals as diverse as humans, insects and nematode worms. Fundamental to these processes is the abundance of information a cell receives from other cells and tissues in its neighbourhood. This information is received as chemical messages which may be in the form of relatively simple molecules, such as inorganic ions, hormones or transmitters, or by contact with proteins on the surface of other cells.

Table 1.1 provides a list of completed whole-genome DNA sequences as at January 1998. Many more bacterial pathogens and organisms that are human parasites are being studied. Table 1.2 lists some genome sequencing projects that are in progress. It is very difficult to interpret human sequences and it is expected that conserved regions in common with the mouse will help in under-

**Table 1.1**   Completed whole-genome sequences.

|  | Size (Mb) |
|---|---|
| Archaea |  |
| *Archaeoglobus fulgidus* | 2.1 |
| *Methanobacterium thermoautotrophicum* | 1.8 |
| *Methanococcus jannaschii* | 1.7 |
| *Pyrococcus horikoshii* | 2.0 |
| Eubacteria |  |
| *Bacillus subtilis* | 4.2 |
| *Borrelia burgdorferi* | 1.2 |
| *Escherichia coli* | 4.6 |
| *Haemophilus influenzae* | 1.8 |
| *Helicobacter pylori* | 1.7 |
| *Mycobacterium tuberculosis* | 4.4 |
| *Mycoplasma genitalium* | 0.6 |
| *Mycoplasma pneumoniae* | 0.8 |
| *Synechocystis* sp. PCC6803 | 3.6 |
| *Treponema pallidum* | 1.1 |
| Fungi |  |
| *Saccharomyces cerevisiae* | 13.5 |

**Table 1.2**   Some genomes that are being sequenced.

|  |  | Size (Mb) |
|---|---|---|
| Fission yeast | *Schizosaccharomyces pombe* | 14 |
| Thale cress | *Arabidopsis thaliana* | 70 |
| Malaria parasite | *Plasmodium falciparium* | 30 |
| Nematode worm | *Caenorhabditis elegans* | 100 |
| Fruit fly | *Drosophila melanogaster* | 165 |
| Puffer fish | *Fugu rubripes* | 400 |
| Mouse | *Mus musculus* | 3000 |
| Man | *Homo sapiens* | 3000 |

standing gene structures. The puffer fish has a compact vertebrate genome that will also help to elucidate gene structure and exon/intron boundaries. The vertebrate genomes are expected to contain about 60 000 genes. The mouse is essential as a mammal for experimental work.

To understand the content of genomes it is essential to have an appreciation of their history. Since ancient genomes cannot be directly observed (except in a few cases of fossils that yield DNA sequences) the history has to be inferred by comparative studies. The subject of sequence similarity determination is treated in detail in Chapter 6 and information about families of sequences is found in Chapters 9–11. Comparative sequence studies start with database searches which allow a newly determined sequence to be compared to the universe of known sequences. The most popular programs to do this are called BLAST and FASTA. Once similarities have been detected, more sophisticated methods are used to place the sequence in the accurate context of others.

Similar sequences may have descended from a common ancestral sequence, in which case it is inferred that they are homologous. No proof of homology exists because it is an hypothesis not a fact. The concept is extended further to cover the cases of sequence evolution in a lineage of organisms or sequence evolution within a genome. As organisms speciate and become reproductively isolated, they may have variants of a protein that still perform the same function but differ in sequence. As divergence proceeds, the proteins may adopt different functions but still share a common ancestor. Homologous sequences that have evolved in different lineages are called orthologues. Alternatively, sequences may duplicate and diverge within a genome, in which case they are called paralogues. Whole genomes may sometimes duplicate and the genes undergo divergence and selective deletion. These processes all go on together to generate complex gene families and it may be a very difficult task to unravel their histories.

The number of genes in vertebrate animals is thought to be the results of two rounds of genome duplication. Amphioxus may contain 20 000 genes, the lamprey and hag fish 40 000 genes, and amphibians, reptiles, birds and mammals 60 000 genes. This needs to be confirmed by further experimental work. These evolutionary considerations are further elaborated in Chapter 11.

The evolutionary divergence of modern life forms from common ancestors provides a marvellous unifying principle to understand extant informational macromolecules. Nature (via natural selection) is happy to adopt things that work for a purpose and has no interest in history. We must therefore be careful not to be misled by cases of different function from the same ancestral sequence or, alternatively, cases of the same function from different ancestral sequences. As is emphasized later in the book, functional domains of molecules are the important unit, not the whole protein, and may be reshuffled.

## 1.3  PHENOTYPES AND GENOTYPES

Individuals within a species vary and this variation can be ascribed both to variations in the environment and to their genetic make up. Genetic variations have been exploited since the dawn of agriculture in programmes of selective breeding of animals and plants. With increased control of viral, bacterial and parasitic infections, medical science has focused on diseases of genetic origin. Single gene defects are relatively rare in the population and many have been well characterized in terms of their underlying molecular basis as, for example, in cystic fibrosis. Conditions that are correlated to the sequences of numbers of genes are common and include diabetes, arthritis and heart disease. The term cancer covers a variety of genetic defects that may be both hereditary and somatic, developing in body cells of an individual. Agriculture and medicine are being revolutionized by the understanding provided by molecular biology and this depends on the connection of observable characteristics (phenotypes) to responsible DNA sequences (genotypes). The ability to replace a parentally derived sequence by a modified sequence in an experimental animal such as a mouse can provide living proof of sequence function.

Genetic linkage analysis provides the means by which phenotypes may be connected to genotypes (Chapter 3). Linkage analysis requires two or more variants of each of two or more observable characteristics of an organism. Some characteristics are not inherited together by more than the chance amount (50:50), while others may show varying degrees of co-inheritance. By suitably designed breeding experiments, the degree of this linkage of characteristics may be measured. For example, morphological characters of fruit flies (*Drosophila*) could be placed in linear orders in a number of distinct linkage groups and it was realized that these corresponded to the chromosomes observable by microscopy.

The characters studied in linkage analysis need not be morphological. In recent years detailed linkage maps of both human and mouse have been built up using a variety of highly polymorphic molecular markers (Chapter 3). The human map has limited resolution for practical reasons that include the generation time and ethical considerations. The mouse map can be refined to a resolution of a few hundred kilobases (kb), which is enough to connect a phenotype to a single gene in many cases. It thus provides a remarkably powerful resource for mammalian biology.

Linkage maps are also being established for a variety of species of agricultural importance, both animals and plants. In mammals, fewer than 40 chromosome break-points may be required to convert one karyotype into another, with cattle, pigs, sheep and goats being considerably more similar. Within plant families such as Brassicae, Solanaceae, and Graminae there is very good conservation of linkage that can be applied to horticulture.

Local gene orders (synteny) tend to be conserved because chromosome rearrangements involve relatively large segments of chromosomes. Comparative linkage studies can help to confirm ideas about gene family relationships that have been derived from sequence comparisons (orthology and paralogy).

## 1.4  PHYSICAL MAPPING

The concern of the past few years has been whole-genome sequencing. It is now routine for viral and bacterial sequences to be sequenced by a whole-genome shotgun procedure that breaks the genome into random overlapping fragments that can be cloned and sequenced directly. While it has recently been suggested that such a procedure should work for the human genome, it is not clear that this will be the case because of repetitive sequence families that have very similar members of over 1000 bases long (Chapter 7).

Physical mapping is the process of characterizing the DNA so that it can be handled in less than genome-sized chunks. The physical map with perhaps the lowest resolution, but nevertheless very effective for many purposes, is the cytogenetic map obtained by visualizing the affinity of a molecular marker to a chromosome spread. The development of chromosome-specific paints represents the culmination of cytogenetic technique.

A physical map with greater resolution is a radiation hybrid map. To produce this, cells are irradiated to break up the DNA, the size of the chunks being related to the dose given. Then, some of the DNA fragments are rescued at random into a rodent cell. A cell line made from this material can be characterized in terms of the DNA fragments rescued and a set of previously mapped molecular markers. With a panel of about 100 cells and several thousand markers, one can start to map the likely chromosome position of unknown probes. The process is somewhat akin to linkage analysis in that a likelihood is computed, but differs in being able to map single probes that need not be polymorphic.

Another type of physical map is an overlapping clone map that is constructed from shotgun fragments of DNA that have been cloned into a suitable vector. The map may be made of a local region or of a whole genome. Cosmid clones can accommodate about 50 kb of DNA, yeast artificial chromosome (YAC) clones over 1000 kb and bacterial artificial chromosome (BAC) clones about 200 kb. Cosmid libraries of single human chromosomes exist but suffer to some extent from deletion and contamination with DNA from other chromosomes. Cosmids are unsuitable to contain the whole human genome as too many clones would be required. For this reason, YAC clones were used to attempt to make an overlapping clone map for the human genome. This was only partly successful because YAC clones are rather unstable and tend to recombine so that the DNA within them comes from two or more chromosomes. The process is facilitated by the numerous repeated sequences in the human genome. Sequence-ready clone maps are now being prepared using BAC clones that are of a size large enough to cover the genome in a reasonable number of clones. They offer the advantage that DNA from a BAC can be shotgunned and subcloned into M13 for direct sequencing.

To construct an overlapping clone map, each clone is characterized in terms of a restriction fingerprint, in terms of hybridization to a number of molecular probes, or by polymerase chain reaction (PCR) with a variety of primer pairs (sequence tagged sites, STSs). Clones sharing markers are then assembled into a single contiguous DNA strand (contig).

An important type of molecular marker is a cDNA clone that has been characterized from a cDNA library made from expressed RNA messages. Many cDNA clones have been sequenced from a variety or organisms and cells/tissues because it is very hard to predict which DNA is coding (see Chapters 4 and 8). If a clone comes from a cDNA library there is a good chance (although not a certainty) that it represents transcribed material. When primer pairs are designed from sequenced cDNAs these are known as expressed sequence tags (ESTs). A detailed transcript map of the human genome is being prepared using ESTs and radiation hybrid mapping and this will be very valuable in interpreting the genomic sequence.

Prediction of genes in bacteria, yeast and nematode by the use of a computer is relatively successful. The same cannot be said for mammalian genes. Comparative sequence analysis is more successful in revealing conserved regions that represent not only DNA coding for protein but also control regions such as promoters and enhancers. Because rates of divergence of genes (lack of conservation) vary, very useful information is being obtained by comparing homologous regions of the human, mouse and fugu genomes that differ in estimated divergence time by an order of magnitude. Comparative sequence analysis is very efficient in limiting the number of biological experiments that need to be performed to elucidate DNA sequence function.

## 1.5    EXPRESSION PROFILING

The current availability of hundreds of thousands of cDNA sequences that represent new putative transcripts with encoded proteins presents a considerable challenge in terms of elucidating their function. The cell is the fundamental unit in which expression should be studied. In the nematode, the lineage of about 1000 cells forming the adult organism has been worked out in detail. Knowledge of the localization of proteins within the cells will complete the static picture of the details of gene expression.

In other organisms attempts are also being made to map gene expression (Chapter 14). For example, the anatomy of the embryonic mouse at various stages of development has been reconstructed from serial sections in a computer model. Expression patterns of specific cDNAs within the embryo can then be recorded from hybridization experiments. This approach allows one to deal with one transcript at a time.

In order to deal with many thousands of transcripts simultaneously, a different approach is needed involving a high throughput methodology. For humans, one such approach is to use cell lines that represent a particular cell type. Transcripts from the cell line are converted to cDNA and cloned in a suitable vector. The cDNA clones are then amplified by PCR and separated on gels. Since too many products would be produced from the entire population, these may be amplified in subsets by cutting with a restriction enzyme, ligating with an adapter to sticky ends and then performing the PCR in, for example, 64 batches each with different dinucleotides in the primers. Known genes can be identified from the products obtained, while unknown genes can be characterized by sequencing products that do not appear to correspond to previously characterized fragments. In another approach, oligonucleotide microarrays (so called DNA chips) may be employed for high-throughput detection of gene expression.

Alternatively, if one is interested in a particular gene family, primers can be designed that amplify the family members and these primers can be applied to the study of a large number of different cell types.

These, and similar methods can give a large amount of information very quickly about the genes that are expressed in cells of known origin. Analysis of these data may reveal interesting patterns. This may provide a clue as to function, but further work is needed to discover how the genes act together.

## 1.6   MULTIPROTEIN COMPLEXES AND PATHWAYS

Many important cellular processes are performed and regulated by multiprotein complexes such as those involved in transcription, translation, receptors and cytoskeletal structures. The function of many of these proteins would be entirely mysterious if studied in isolation. Other proteins are participants in pathways and interact with a small number of other proteins but do form an essential link in a long chain of relationships. Methods have recently been developed that permit investigation of these proteins and their identification against putative proteins from genomic sequencing or EST databases.

Protein complexes may be isolated from cells under carefully controlled conditions and purified away from contaminants. The protein complexes may then be dissociated and the component protein separated by two-dimensional gel electrophoresis. Tryptic digests of excised protein spots may be characterized by mass spectrometry as peptide sequences. Search of EST databases (particularly the 5' clones) permits the correspondence between a protein spot and an EST sequence to be established. These techniques have been established in yeast where the total genomic sequence is known. It appears that sufficient ESTs have been sequenced for the approach to be largely successful also in mammals. This outline description hides considerable technical complexity;

for example, if a target protein is known to be a participant in a complex of interest it may be tagged to facilitate the isolation of the complex.

Other techniques that are becoming widely used are the yeast one-hybrid and two-hybrid system for identifying DNA-binding proteins and protein–protein interactions, respectively. The yeast transcription activator protein Ga14p has a DNA-binding domain and an activation domain that may be separated. Transcriptional activation is detected when the binding domain is bound to the DNA recognition sequence and the activation domain is attached nearby. The two-hybrid system involves fusing the Ga14p-binding domain with a first protein and the activating domain with a second protein. If the two proteins interact then transcriptional activation is restored and can be reported with a gene such as *LacZ*. The method can be used to screen a protein of interest against a variety of possible targets that might interact with it.

## 1.7  SEQUENCE, STRUCTURE AND FUNCTION

Given a protein sequence, is it possible to predict its structure and from this predict its function? Such a dream has been implicit in much of the literature for the past 50 years. The dream will remain such and the general problem is insoluble (Chapter 12). In reality, each element of the triple is open to experimental solution. Finding out what a protein does can be tackled in isolation from a knowledge of its sequence and structure. Protein folding is too complex and subtle to predict from sequence alone.

It has become apparent that it is necessary to document many folds and functions but that the comparative method in biology founded upon evolutionary principles rescues us from the mire of complexity. Much remains to be done to accumulate the necessary knowledge, but there are constraints on folds and functions that give us confidence that one day the task will be completed.

# 2 Nucleic Acid and Protein Sequence Databases

*Gary Williams*

HGMP Resource Centre, Hinxton, Cambridge, UK

## 2.1 INTRODUCTION

The most common uses of the sequence databases are to search for similarity with an unknown query sequence and to search for entries matching keywords in their annotation.

You may already be familiar with using BLAST or FASTA which report alignments of regions of similarity between database entries and your unknown sequence, or with using SRS or Entrez which allow you to find database entries by keyword searches of their annotation.

This chapter attempts to introduce the databases so that some of the details of the information they hold will become clearer and potential problems associated with them can be highlighted and avoided.

## 2.2 THE MAIN SEQUENCE DATABASES

There are two main nucleic acid sequence databases and one main protein sequence database in widespread general use amongst the biological community. For nucleic acid these are EMBL (Stoesser *et al.*, 1998) and Genbank (Benson *et al.*, 1998) and for protein this is SWISS-PROT (Bairoch & Apweiler, 1997).

There are also many databases that contain special purpose sets of sequences, subsets of sequences derived from the main ones, databases of complete genomes, databases of secondary structure or other derived or additional information and unpublished, private or commercially available sequence databases. Most of these will not be discussed in this chapter but some examples can be found in Table 2.1.

*Genetics Databases*
ISBN 0-12-101625-0

**Table 2.1**   Miscellaneous databases.

VectorDB
http://vectordb.atcg.com/
Contains the sequences for many vectors, all in Genbank format.

Codon Usage Database
http://www.dna.affrc.go.jp/~nakamura/codon.html
Codon Usage Database is an extended WWW version of CUTG (Codon Usage
Tabulated from GenBank). The frequency of codon use in each organism may be searched
through this site.

ImMunoGeneTics database (IMGT)
http://imgt.cnusc.fr:8104/
An integrated database specializing in immunoglobulins (Ig), T-cell receptors (TcR) and
major histocompatibility complex (MHC) molecules of all species.

The Transcription Factor Database (TRANSFAC)
http://transfac.gbf.de/TRANSFAC/
A database of eukaryotic *cis*-acting regulatory DNA elements and *trans*-acting factors. It
covers the whole range of species from yeast to human.

The Tumour Gene Database
http://condor.bcm.tmc.edu/oncogene.html
Contains information about genes which are targets for cancer-causing mutations; proto-
oncogenes and tumour-suppressor genes. Its goal is to provide a standard set of facts (e.g.
protein size, biochemical activity, chromosomal location, etc.).

The RNA World
http://www.imb-jena.de/RNA.html
The RNA World at IMB Jena holds links to many small RNA databases and RNA related
topics.

Nucleic Acid Database (NDB)
http://ndb-mirror-2.rutgers.edu/
The goal of the Nucleic Acid Database Project (NDB) is to assemble and distribute
structural information about nucleic acids.

Dictionary of protein sites and patterns (PROSITE)
http://www.expasy.eh/prosite/
A database of biologically significant sites, patterns and profiles in proteins that help to
identify reliably to which known family of protein (if any) a new sequence belongs.

Protein Motif Fingerprint Database (PRINTS)
http://www.biochem.ucl.ac.uk/bsm/dbbrowser/PRINTS/PRINTS.html
A database of groups of conserved motifs used to characterize a protein family, they can
encode protein folds and functionality more flexibly and powerfully than can single
motifs.

Protein domain families (Pfam)
http://www.sanger.ac.uk/Software/Pfam/
A high-quality comprehensive collection of protein domain families.

Protein domain database (ProDom)
http://protein.toulouse.inra.fr/prodom.html
A protein domain database produced automatically from the SWISS-PROT database.

Annotated protein domain sequences (SBASE)
http://base.icgeb.trieste.it/sbase/
A database of annotated protein domains.

Genome Sequence DataBase (GSDB)
http://www.ncgr.org/gsdb/
A comprehensive DNA database, this is a competitor with the NCBI for the Genbank contract.

G-Protein Coupled Receptor Database (GCRDb)
http://www.gcrdb.uthscsa.edu/
Holds sequence data of the G protein-coupled receptor class of proteins.

The Restriction Enzyme Database (REBASE)
http://rebase.net.com/
A collection of information about restriction enzymes, methylases, the micro-organisms from which they have been isolated, recognition sequences, cleavage sites, methylation specificity, the commercial availability of the enzymes and references.

Immunological Database (Kabat)
http://immuno.bme.nwu.edu/
A database of published immunological protein sequences.

NRL_3D
http://www.rcsb.org/pdb/
The sequences and annotation of the proteins in the Brookhaven Protein Databank (PDB) of crystallographic structures.

O-GlycBase
http://www.cbs.dtu.dk/databases/OGLYCBASE/
A database of proteins with at least one experimentally verified O-glycosylation site.

dbEST
http://www.ncbi.nlm.nih.gov/dbEST/
Sequence and mapping data on 'single-pass' cDNA sequences or Expressed Sequence Tags from a number of organisms.

dbSTS
http://www.ncbi.nlm.nih.gov/dbSTS/
Sequence and mapping data on short genomic landmark sequences or sequence tagged sites.

UniGene
http://www.ncbi.nlm.nih.gov/UniGene/
Clusters of human EST sequences that represent the transcription products of distinct genes.

**Table 2.1** Continued.

---

V BASE
http://www.mrc-cpe.cam.ac.uk/imt-doc/public/INTRO.html
A directory of human immunoglobulin germ line variable region sequences.

Human CpG Island database
http://biomaster.uio.no/cpgisle.html
A database of short and dispersed regions of unmethylated DNA with a high frequency of CpG dinucleotides relative to the bulk genome.

Human Gene Mutation Database
http://www.uwcm.ac.uk/uwcm/mg/hgmd0.html
A collation of the majority of known (published) gene lesions responsible for human inherited disease.

Genome Sequencing Projects
http://www.mcs.anl.gov/home/gaasterl/genomes.html
An up-to-date list of genome sequencing projects with links to genome sites and progress reports.

---

## 2.2.1    The main nucleic acid databases

The European Bioinformatics Institute (EBI, http://www.ebi.ac.uk/), the National Center for Biotechnology Information (NCBI, http://www.ncbi.nlm.nih.gov/) and the DNA Data Bank of Japan (DDBJ, http://www.ddbj.nig.ac.jp/) collaborate to produce the nucleic acid databases EMBL, Genbank and DDBJ.

The collaboration includes a common feature table (http://www.ebi.ac.uk/ebi_docs/embl_db/ft/feature_table.html) which describes biologically interesting regions in the sequence entries, a common set of unique sequence identification numbers and rapid exchange of submitted sequences.

The exchange of sequences occurs daily, so that each of the three main databases holds the same data. The sequence data is exactly the same in each database. There may be very slight and largely insignificant differences in the annotation of the entries and, as we shall see, there are stylistic details in how the databases present the information.

So why are there three databases and not one? The sequence databases represent a valuable commercial resource for any organization interested in biotechnology or pharmaceuticals. The health research funding bodies in the USA, Europe and Japan each started their own database organization as a strategic commercial resource and these have continued as separate but cooperating organizations.

It is common for database searching systems such as the Entrez or BLAST WWW sites to merge different databases into one 'nonredundant' database. This is done by taking one of the major databases, for example GenBank, and then adding in unique sequences from some of the smaller, less significant nucleic acid databases. Thus, when you find yourself searching a database that calls itself 'nr', you are searching a nonredundant database that is predominantly derived from the EMBL or GenBank database.

## 2.2.1.1   What they hold

EMBL and Genbank are historical archives or databases of records of all publicly available DNA sequences. That is to say, they contain a record purely of what has been placed in them with no intention to ultimately merge overlapping or duplicated sequences into contiguous regions. The result of this is that there can be many nearly identical versions of the same sequence in the databases.

An entry in the databases can contain anything from an anonymous partial cDNA, to a single exon, to a complete mRNA sequence, to the genomic sequence of a gene, to complete cosmids or even larger sequences covering several genes. The longest sequences in EMBL and GenBank are split at the arbitrary size of 350 kb to enable computer software to manipulate them more easily.

The quality of the annotation of the entry can range from the bare details of the organization that produced it and the tissue and species it is derived from, to a very detailed description of every feature in the sequence with many cross references to related entries in other databases.

## 2.2.1.2   How the data is entered

Sequences are placed in the databases from published papers describing them, or more commonly, they are submitted directly by their authors to the database organizations. The sequence is then deemed to belong to the author and only they can update or amend it. This may cause problems with continued curatorship of these data as authors leave the profession or lose interest in the sequence.

Most journals require submission of sequence information to the sequence databases prior to publication. There is therefore an incentive for authors to submit their sequences to the databases as quickly as possible. The preferred method of submission is by the WWW.

Webin (http://www.ebi.ac.uk/submission/webin.html) is the WWW site for submitting nucleotide sequence data and associated biological information to the EMBL database at the EBI.

BankIt (http://www.ncbi.nlm.nih.gov/BankIt/) is the NCBI equivalent WWW site for submitting to GenBank.

Sequin (http://www.ncbi.nlm.nih.gov/Sequin/) is a program which can be downloaded and run on the authors local computer for preparing a sequence for submission. The result is then sent by e-mail to the NCBI or the EBI.

It is only necessary to submit a sequence to one of the databases. Submissions to any of the three nucleic acid databases are forwarded daily for inclusion in the other two. When the submission is received, database staff will provide the author with a unique database accession number to identify the sequence. This accession number should be used in references to the sequence in any publication.

### 2.2.1.3   Who updates the data

The sequences can normally only be altered by the authors. If you think you have discovered an error in a sequence, you must contact the database who will in turn contact the authors.

### 2.2.1.4   Example entry

Figure 2.1 gives examples of the same EST sequence entry taken from the EMBL and the GenBank databases. You will note that the entry is composed of annotation and sequence data. The textual annotation is split into fields with a key on the left hand side. The key is composed of two letters in EMBL, in Genbank it is a word. The sequence is placed at the end of the entry in both the databases. Except for the stylistic details, the information in these two entries is substantially the same.

### 2.2.1.5   Description of the fields

The annotation fields contain the following main sets of information:

- *ID/LOCUS:* this line holds several items of information. First, an entry name such as 'AA184' or 'R47184', which is only of use in identifying this entry in this version of this database, i.e. it should be used sparingly, although some people favour it. Note that the entry name is not the same between these two databases. The second item is the review status of the sequence. The third item is the type of molecule (e.g. RNA or DNA). The fourth item is the taxonomic division (see below) within the EMBL or GenBank database that the entry is assigned to, and the last item is the sequence length.

**(a)**

```
ID   AA184     standard; RNA; EST; 201 BP.
XX
AC   R47184;
XX
NI   g807526
XX
DT   18-MAY-1995 (Rel. 43, Created)
DT   24-AUG-1995 (Rel. 44, Last updated, Version 3)
XX
DE   LF168T7 Aedes aegypti cDNA clone LF168 5'.
XX
KW   EST.
XX
OS   Aedes aegypti (yellow fever mosquito)
OC   Eukaryotae; mitochondrial eukaryotes; Metazoa; Arthropoda; Tracheata;
OC   Hexapoda; Insecta; Pterygota; Diptera; Nematocera; Culicoidea;
OC   Culicidae; Aedes.
XX
RN   [1]
RP   1-201
RA   Severson D.W., Mori A., Zhang Y., Christensen B.M.;
RT   "Linkage map for Aedes aegypti using restriction fragment length
RT   polymorphisms";
RL   J. Hered. 84:241-247(1993).
XX
CC   Other_ESTs: LF168SP6 Map: Chromosome Contact: Severson DW Vector
CC   Biology University of Wisconsin Department of Animal Health and
CC   Biomedical Sciences, 1655 Linden Drive, University of Wisconsin,
CC   Madison, WI 53706 Tel: 6082652647 Fax: 6082627420 Email:
CC   dave@aedes.ahabs.wisc.edu. NCBI gi: 807526
XX
FH   Key          Location/Qualifiers
FH
FT   source       1..201
FT                /organism="Aedes aegypti"
FT                /note="yellow fever mosquito"
FT                /clone="LF168"
FT                /strain="Liverpool"
FT   mRNA         <1..>201
FT                /map="3"
XX
SQ   Sequence 201 BP; 47 A; 60 C; 49 G; 42 T; 3 other;
     ggaggacgct gcaagcacaa ccncggccac gtcaaggccg tgcnttgcac caattgcgcc      60
     cgctgcgtcc cgaaggataa ggcgatcaag aagttcgtga tccggaacat cgtcgaagcc     120
     gctgccgttc gagatatttc ggatgcttcc tgctacaact cctacgttct ncccaaactt     180
     tacgaaaatt tgcactactg g
```

**Figure 2.1**   (a) A typical EMBL entry.

**(b)**

```
LOCUS          R47184     201 bp  mRNA       EST     15-MAY-1995
DEFINITION     LF168T7 LIV female Aedes aegypti cDNA clone LF168 5'.
ACCESSION      R47184
NID            g807526
KEYWORDS       EST.
SOURCE         yellow fever mosquito.
  ORGANISM     Aedes aegypti
               Eukaryotae; Metazoa; Arthropoda; Tracheata; Hexapoda; Insecta;
               Pterygota; Diptera; Nematocera; Culicoidea; Culicidae; Aedes.
REFERENCE      1 (bases 1 to 201)
  AUTHORS      Severson,D.W., Mori,A., Zhang,Y. and Christensen,B.M.
  TITLE        Linkage map for Aedes aegypti using restriction fragment length
               polymorphisms
  JOURNAL      J. Hered. 84, 241-247 (1993)
  MEDLINE      93340446
COMMENT        Other_ESTs: LF168SP6
               Contact: Severson DW
               Vector Biology
               University of Wisconsin
               Department of Animal Health and Biomedical Sciences, 1655 Linden
               Drive, University of Wisconsin, Madison, WI 53706
               Tel: 6082652647
               Fax: 6082627420
               Email: dave@aedes.ahabs.wisc.edu
               Seq primer: T7.
FEATURES             Location/Qualifiers
     source          1..201
                     /organism="Aedes aegypti"
                     /strain="Liverpool"
                     /note="Vector: lamdaGEM4; Site_1: EcoRI; Site_2: XbaIcDNA
                     was directionally synthesized from the XbaI site in the
                     vector to the EcoRI site."
                     /db_xref="taxon:7159"
                     /map="3"
                     /clone="LF168"
                     /clone_lib="LIV female"
BASE COUNT     47 a    60 c    49 g    42 t    3 others
ORIGIN
        1 ggaggacgct gcaagcacaa ccncggccac gtcaaggccg tgcnttgcac caattgcgcc
       61 cgctgcgtcc cgaaggataa ggcgatcaag aagttcgtga tccggaacat cgtcgaagcc
      121 gctgccgttc gagatatttc ggatgcttcc tgctacaact cctacgttct ncccaaactt
      181 tacgaaaatt tgcactactg g
```

**Figure 2.1** *(continued)* (b) The same entry from the GenBank database.

■ *AC/ACCESSION (e.g. R47184):* this unique identifier is guaranteed to stay with and identify this sequence for the rest of its life and is the same between all the databases holding this sequence. This is the name that should generally be used to refer to this sequence in the literature. When sequences from different entries are merged a new accession number will be created for the new sequence. However, the old accession numbers will also be quoted in this field as secondary accession numbers, allowing continued identification of the constituent sequences. Until about 1997 the accession numbers were composed of one letter followed by five digits (e.g. A12345) but numbers assigned since 1997 are composed of two letters followed by six digits (e.g. AB123456).

■ *NI/NID (e.g. g807526, an internal database key):* this is being phased out in favour of a new system of identifiers for both nucleotide and protein sequences, of the form 'Accession.Version' (e.g., AB000349.3). The accession portion of these identifiers is stable and will not change, but the version portion will be incremented whenever the underlying sequence changes. This will allow identification of a specific version of the sequence.

■ *DE/DEFINITION:* a single line containing a clear and concise description of the entry. This is what is commonly reported when a search finds the entry. The description of anonymous EST entries should ideally contain the best effort to give a short description of the sequence based on homology to families of known sequences. Such description lines are bound to become outdated and may be misleading, but may still be informative.

■ *KW/KEYWORDS (e.g. EST):* a dry list of keywords that may or may not be informative. The information in this field has often already been given in the DEFINITION field.

■ *OS,OC/SOURCE,ORGANISM:* the taxonomic names of the organism from which the sequence was determined.

■ *RN,RA,RT,RL/REFERENCE,AUTHORS,TITLE,JOURNAL,MEDLINE:* one or more articles in journals describing the sequence, its origin, analysis and significance.

The RN/REFERENCE line gives a unique number to each reference citation within an entry.

The RA/AUTHORS line lists the authors of the paper.

The RT/TITLE line gives the title of the paper.

The RL/JOURNAL line contains the conventional citation information for the reference.

The MEDLINE line gives an ID number for the Medline bibliographic database.

- CC/COMMENT: one or more other sets of information that do not
  fit neatly into the above sets of fields.

- FH,FT/FEATURES: features of the sequence, see below for a brief
  description of feature tables.

## 2.2.1.6   Feature tables

The EMBL, GenBank and DDBJ databases share a common feature table format. The purpose of the feature table is to show information on biologically
meaningful features in the sequence entry and to indicate variants of the
sequence.

The layout of this table is a textual key on the left with location information
on the right, optionally followed by qualifiers (starting with a solidus) which
supply additional information about the feature that is not given by the key
and location. See Fig. 2.2 for an example.

There are many keys defined for use in the feature table, see Table 2.2 for a
list of the more commonly used ones.

The location descriptors can specify many types of sites. The most common
of these include: a single base number (e.g. '88'); a range of sequence where the
start and end positions are separated by two dots (e.g. 76..876); and a site
between two adjacent bases: the two positions are separated by a caret (e.g.
44^45). Uncertainty about a starting or ending point of a location can be indicated by the < or > symbols (e.g. <78..1034 means that the range starts somewhere before position 78).

Operators can specify the reverse strand (e.g. complement(76..876)), a
group of locations that are ordered but not necessarily joined (e.g. order
(45..63,67..90)) or a contiguously joined set of sequences (e.g. join
(33..56,102..223,389..1293)).

Most of the qualifiers of features are fairly self-explanatory, but the following may need some clarification:

- /citation=[1]: a reference to a citation listed in the entry's annotation.
- /db_xref=SWISS-PROT:P12345: a cross-reference to an entry in
  another database holding related information.
- /evidence=experimental: the nature of supporting evidence for this
  feature.
- /gene=Fau: symbol of the gene corresponding to a sequence region.
- /map=6p1-2: genomic map position of the feature. For human
  sequences this takes the form of a cytological location for example,

6p1-2 means in bands 1 to 2 of the short arm of chromosome 6. It is common for the entry to only have been mapped to a chromosome, e.g. 11.

- */protein_id=AAA12345.1:* a cross-reference to an entry in the SWISS-PROT protein database giving the accession number and version of the protein entry.

See   http://www.ebi.ac.uk/ebi_docs/embl_db/ft/feature_table.html   for   full details of the feature table definition.

```
Key     Location/Qualifiers

source  1..4276
        /organism="Arabidopsis thaliana"
        /strain="Landsberg"
        /db_xref="taxon:3702"
mRNA    join(<1264..1402,2497..4235)
        /gene="PT2"
gene    <1264..4235
        /gene="PT2"
exon    <1264..1402
        /gene="PT2"
intron  1403..2496
        /gene="PT2"
exon    2497..4235
        /gene="PT2"
CDS     2509..4113
        /gene="PT2"
        /codon_start=1
        /product="phosphate transporter"
        /db_xref="PID:g2564661"
        /translation="MAREQLQVLNALDVAKTQWYHFTAIIIAGMGFFTDAYDLFCISL
        VTKLLGRIYYHVEGAQKPGTLPPNVAAAVNGVAFCGTLAGQLFFGWLGDKLGRKKVYG
        MTLMVMVLCSIASGLSFGHEPKAVMATLCFFRFWLGFGIGGDYPLSATIMSEYANKKT
        RGAFVSAVFAMQGFGIMAGGIFAIIISSAFEAKFPSPAYADDALGSTIPQADLVWRII
        LMAGAIPAAMTYYSRSKMPETARYTALVAKDAKQAASDMSKVLQVEIEPEQQKLEEIS
        KEKSKAFGLFSKEFMSRHGLHLLGTTSTWFLLDIAFYSQNLFQKDIFSAIGWIPPAQS
        MNAIQEVFKIARAQTLIALCSTVPGYWFTVAFIDVIGRFAIQMMGFFFMTVFMFALAI
        PYNHWTHKENRIGFVIMYSLTFFFANFGPNATTFVVPAEIFPARFRSTCHGISAASGK
        LGAMVGAFGFLYLAQNPDKDKTDAGYPPGIGVRNSLIVLGVVNFLGILFTFLVPESKG
        KSLEEMSGENEDNENSNNDSRTVPIV"
```

**Figure 2.2**   An example feature table from entry AF022872 (*Arabidopsis thaliana* phosphate transporter (PT2) gene, complete cds).

**Table 2.2** Some examples of commonly used keys in the nucleic acid sequence feature table include:

| Key | Description |
|---|---|
| CDS | Coding sequence for a protein including stop codon |
| exon | Region of genome that codes for spliced mRNA |
| gene | Region identified as a gene for which a name has been assigned |
| intron | Transcribed DNA spliced out of mature mRNA |
| mat_peptide | Sequence for the protein following post-translational modification |
| misc_binding | Site in nucleic acid which binds another moiety |
| misc_feature | Region which cannot be described by any other feature key |
| misc_signal | Region containing a novel signal for gene function or expression |
| mRNA | Messenger RNA; includes 5'UTR, CDS, exon and 3'UTR |
| polyA_signal | Recognition region that is followed by polyadenylation |
| polyA_site | Site on an RNA transcript which will receive a poly-A tail |
| prim_transcript | Primary, unprocessed transcript including 5'clip, introns, 3'clip |
| promoter | Region of a DNA molecule involved in RNA polymerase binding |
| repeat_region | Region of genome containing repeating units |
| repeat_unit | Single repeat element |
| rRNA | Mature ribosomal RNA |
| sig_peptide | Coding sequence for an N-terminal domain of a secreted protein |
| source | Identifies the biological source of the specified span of the sequence. This key is mandatory. Every entry will have, as a minimum, a single source key spanning the entire sequence. More than one source key per sequence is permissible |

## 2.2.1.7 The database divisions

The EMBL, GenBank and DDBJ databases are divided into sections based either on taxonomy or function (Ouellette & Boguski, 1997). Many sites providing keyword or sequence similarity searches of these databases offer facilities to either search the complete database or any of the divisions of that database. This allows searches and analyses of sequences in the databases to be targeted at a relatively narrow group of species.

In addition to the formal divisions, many sites will further split the sequences into subdivisions that can be analysed locally. For example, it is common to be able to search EST sequence subdivisions that have been created to hold just human, mouse or other species EST sequences. It is also common to hold local databases of the complete genomes of selected organisms.

Not all of the divisions occur in all three databases, some of the databases group some of the species differently so, for example, EMBL holds all of the human sequences in the 'Human' division and other primates are held in the 'Mammal' section, while GenBank and DDBJ hold all the primate sequences, including human sequences, in the 'Primate' section.

The divisions are summarized in Table 2.3.

The contents of the taxonomic divisions are largely self-explanatory, i.e. the 'viral' section holds viral sequences. The following functional divisions have been created to hold sequence entries that have a common method of production or a specific use that overrides the utility of grouping them by their taxonomic origin.

**Table 2.3.** The database divisions in EMBL, GenBank and DDBJ[1].

| Division | | Database | |
|---|---|---|---|
| Taxonomic | | | |
| bct | Bacterial | — | GenBank |
| fun | Fungal | EMBL, | — |
| hum | Human | EMBL, | — |
| inv | Invertebrate | EMBL, | GenBank |
| mam | Mammalian | EMBL, | GenBank |
| org | Organelle | EMBL, | — |
| phg | Phage | EMBL, | GenBank |
| pln | Plant | EMBL, | GenBank |
| pri | Primate | — | GenBank |
| pro | Prokaryote | EMBL, | — |
| rna | Structural RNA | — | GenBank |
| rod | Rodent | EMBL, | GenBank |
| syn | Synthetic | EMBL, | GenBank |
| vrl | Viral | EMBL, | GenBank |
| vrt | Other Vertebrate | EMBL, | GenBank |
| unc | Unclassified, Unannotated | EMBL, | GenBank |
| Functional | | | |
| con | Constructed sequences | EMBL, | GenBank |
| est | Expressed Sequence Tag | EMBL, | GenBank |
| gss | Genome Survey Sequences | EMBL, | GenBank |
| htg | High Throughput Genomic | EMBL, | GenBank |
| new | New since last release | EMBL, | GenBank |
| patent | Patent | EMBL, | GenBank |
| sts | Sequence Tagged Site | EMBL, | GenBank |

[1] *(DDBJ has the same divisions as GenBank).*

## 2.2.1.7.1    Constructed sequences (CON)

This division holds information on very long sequences, such as complete genomes and parts of chromosomes.

Entries in the CON division contain no feature table or sequence data. They include information about how the sequence is built from its components, e.g.

CO join(X00123:66..100, X00124:109..1002, gap(2000), X99199:990..10000).

This indicates that three entries should be joined together to make this CON entry. There is a gap, estimated to be 2000 bases long, between the second and third part.

## 2.2.1.7.2    Expressed sequence tag (EST)

The EST division comprises sequences commonly denoted as 'expressed sequence tags' or 'transcribed sequence fragments' or 'partial cDNA', independent of their taxonomic classification. They are poor quality, short (100–600 bp) fragments of expressed sequence.

The primary value of the EST division is to perform sequence similarity searches against using an unknown genomic sequence to demonstrate probable patterns of expression and alternative splicing.

The EST division contains about half of all the sequence data in the databases, with the human and mouse being especially heavily represented. There is a substantial amount of duplication of the sequences in the EST entries and efforts have been made to produce much smaller, nonredundant databases derived by clustering these entries and removing duplication and contamination. Examples include STACK (http://ziggy.sanbi.ac.za/stack/), UniGene (http://www.ncbi.nlm.nih.gov/UniGene/), and the TIGR Database (http://www.tigr.org/tdb/).

EST sequences are usually produced by single-strand, single-read sequencing methods only and are considered to be of lower quality. The error rate is often in the range of 1% and includes insertions and deletions as well as substitutions. You should expect a higher than normal frequency of vector and ribosomal DNA (rDNA) and ribosomal RNA (rRNA) contamination in the EST division (Gonzalez & Sylvester, 1997).

There is some genomic contamination, so you should not expect a similarity between a region of your unknown sequence and an EST sequence to prove conclusively that the region is expressed. It is merely a good indication that this is so and further experimental work will be necessary to demonstrate expression.

Equally, although many of the human and mouse expressed sequences are heavily represented many times over in the EST division, lack of similarity of a region of an unknown sequence to any EST entry is not evidence that the region is not expressed. The coverage is heavily biased in favour of the abundantly expressed genes and is incomplete.

The EST entries generally have only marginal annotation. Do not trust DEFINITION lines in the EST annotation which declare a homology to entries of known function, as these are generally produced by automatic assessment of database search results and are rarely, if ever, updated. They can be informative, but they can also be misleading and out of date. It is best to look in Entrez or dbEST to see the known current similarities between the EST entry of interest and other entries in the databases.

dbEST (http://www.ncbi.nlm.nih.gov/dbEST/) is a database which contains the same entries as the EST division of the sequence databases, but with better presentation of the annotation plus information on the other ESTs obtained from the parent clone, detailed contact information about the contributors, genetic map locations (when available), instructions on obtaining physical DNA clones from IMAGE, the ATCC and other sources and results of sequence similarity searches against the protein and nucleic acid databases.

### 2.2.1.7.3   Genome survey sequences (GSS)

The GSS division is a repository for genomic sequence data that is not appropriate for inclusion in the standard organism-specific divisions. Entries in the GSS division include, but are not limited to, sequence data generated by single pass reads from random genome survey sequences (e.g. Fugu genome survey sequences – see http://fugu.hgmp.mrc.ac.uk/), exon trapped products, Alu PCR sequences and cosmid, BAC, or YAC end clones.

These sequence entries can be useful in mapping and sequencing of larger contigs.

### 2.2.1.7.4   High-throughput genomic (HTG)

The HTG division of EMBL is a set of sequences produced by 'high-throughput genome sequences'. This is composed of very long genomic sequences which are in the process of being produced by high-throughput sequencing projects. The entries are primarily nematode and human. The annotation for many of these records is generated through computer analyses.

This division was created as a result of the 'Bermuda Principles' regarding data produced in publicly funded projects. This is an agreement made between the participants of the first International Strategy Meeting on Human Genome Sequencing meeting at Bermuda, 25th–28th February 1996. The participants included representatives of the Wellcome Trust, the UK Medical Research Council, the NIH NCHGR (National Center for Human Genome Research), the DOE (U.S. Department of Energy), the German Human Genome Programme, the European

Commission, HUGO (Human Genome Organisation) and the Human Genome Project of Japan. (http://www.gene.ucl.ac.uk/hugo/bermuda.htm)

The principles agreed to at this meeting were: 'primary genomic sequence should be in the public domain' and 'primary genomic sequence should be rapidly released.'

In consequence of these principles, sequence assemblies are released as soon as possible; in some centres, assemblies of greater than 1Kb are released automatically on a daily basis and finished annotated sequence are submitted immediately to the public databases.

These sequences are therefore unfinished and do not necessarily represent the final, accurate sequence. Work on the sequence is in progress and the release of this data is based on the understanding that the sequence may change as work continues. The sequence may be contaminated with foreign sequence from *Escherichia coli*, yeast, vector, phage etc. There will be many large gaps of unknown length indicated by regions of Ns in the early phases of producing the sequence entry. The error rate of early stages is high but decreases to the quality of finished sequence (about 1 in 10 000 or less).

When the sequencing has been completed, the entry is moved out of the HTG division and into the division appropriate for that species. Thus, completed nematode sequences are placed in the 'Invertebrate' division.

A sequence similarity search match to a HTG sequence primarily indicates that some other group is carrying out industrial-scale sequencing through an area with similarity to your sequence and you should perhaps contact them to see what location they are working on. If you find an interesting HTG entry, you should frequently check on its progress, as it will be changed to hold more 'finished' and better annotated versions of the sequence.

### 2.2.1.7.5    New since last release (NEW)

This is not, strictly speaking, a division of the databases when they are first released. The NEW section holds those sequences which have been added to the database since the last bimonthly release, have had their annotation changed or have had errors in the sequence corrected by the author. Many sites will set this up as if it were a division of the database they hold. It is worth including a search of this section in addition to a search of the main database release in order to find the latest entries.

### 2.2.1.7.6    Patent

This division aims to incorporate sequences mentioned in patents in the European Patent Office (EPO), the US Patent and Trademark Office and the Japanese Patent Office. The databases include sequences from patents as far back as 1960.

The policy of the EPO is to release data to the public (and to EMBL) 18 months after the patent application date, independent of whether a patent has been granted or not. Immediately after release by the EPO the latest patent sequence data are integrated into the EMBL database and made available to the public.

The NCBI do likewise for patents in the US. One difference is that the US patent office only publishes patent documentation if the patent has been granted.

Patent law is complex and national in scope, so fundamental issues may be resolved differently in different countries. The issue of whether to allow patents on various classes of sequences will no doubt continue for a long time to come.

### 2.2.1.7.7   Sequence tagged site (STS)

The STS section of EMBL is a database of sequences and mapping data of short genomic landmark sequences, or sequence tagged sites, used primarily in physical mapping.

A sequence similarity match to an STS sequence of a marker will help to locate your sequence against the physical map.

dbSTS (http://www.ncbi.nlm.nih.gov/dbSTS/) is a database which contains the same entries as the STS division of the sequence databases, but includes detailed contact information about the contributors, experimental conditions and genetic map locations and results of sequence similarity searches against the protein and nucleic acid databases.

## 2.2.1.8   I.M.A.G.E. clones

The I.M.A.G.E. Consortium (Lennon *et al.*, 1996) (http://www-bio.llnl.gov/ bbrp/image/image.html) was initiated by four academic groups on a collaborative basis after informal discussions led to a common vision of how to achieve an important goal in the study of the human genome: the integrated molecular analysis of genomes and their expression. They share high-quality, arrayed cDNA libraries and place sequence, map, and expression data on the clones in these arrays into the public domain. From this information, they re-array the unique clones to form a 'master array' which they hope will ultimately contain a representative cDNA from each and every gene in the genome. These clones are available free of any royalties and may be used by anyone agreeing with their guidelines.

It is important that orders for clones are placed to I.M.A.G.E. collaborators using the I.M.A.G.E. clone ID. This ID can be obtained via dbEST (http://www.ncbi.nlm.nih.gov/dbEST/). Alternatively, the LENS browser is useful for obtaining I.M.A.G.E. clone IDs. (This site is constantly updated but may not include the most recent clones: http://agave.humgen.upenn.edu/lens/.)

If users wish to translate I.M.A.G.E. Consortium clone IDs to obtain the clone names, they can use the HGMP Translate utility (http://www.hgmp.mrc.ac.uk/BIO/translate.html).

There is often some confusion about what number quoted in the database entries refers to the I.M.A.G.E. clone ID. An I.M.A.G.E. clone ID is currently between five and seven digits long. It has no preceding letter and is found in different places in different databases: in dbEST entries, the I.M.A.G.E. clone ID is the number given in the 'CloneID' field; in GenBank and EMBL entries, the I.M.A.G.E. clone ID is the number on the '/clone=' field in the 'Feature' table. The same number may appear in the source field, following 'human clone=', and on the 'DEFINITION' line following the word clone. Figure 2.3 gives an example of the EMBL entry with the accession number W33426 (I.M.A.G.E. clone ID 352061).

```
ID   MM42612   standard; RNA; EST; 438 BP.
XX
AC   W33426;
XX
NI   g1315352
XX
DT   16-MAY-1996 (Rel. 47, Created)
DT   16-MAY-1996 (Rel. 47, Last updated, Version 1)
XX
DE   mc51f03.r1 Soares mouse embryo NbME13.5 14.5 Mus musculus cDNA
DE   clone 352061 5'.
XX
KW   EST.
XX
OS   Mus musculus (house mouse)
OC   Eukaryotae; mitochondrial eukaryotes; Metazoa; Chordata; Vertebrata;
OC   Mammalia; Eutheria; Rodentia; Sciurognathi; Muridae; Murinae; Mus.
XX
RN   [1]
RP   1-438
RA   Marra M., Hillier L., Allen M., Bowles M., Dietrich N., Dubuque T.,
RA   Geisel S., Kucaba T., Lacy M., Le M., Martin J., Morris M.,
RA   Schellenberg K., Steptoe M., Tan F., Underwood K., Moore B.,
RA   Theising B., Wylie T., Lennon G., Soares B., Wilson R., Waterston R.;
RT   "The WashU-HHMI Mouse EST Project";
RL   Unpublished.
XX
CC   Contact: Marra M/Mouse EST Project WashU-HHMI Mouse EST Project
CC   Washington University School of MedicineP 4444 Forest Park Parkway,
CC   Box 8501, St. Louis, MO 63108 Tel: 314 286 1800 Fax: 314 286 1810
CC   Email: mouseest@watson.wustl.edu This clone is available
```

```
CC   royalty-free through LLNL ; contact the IMAGE Consortium
CC   (info@image.llnl.gov) for further information. MGI:223861 Seq
CC   primer: mob.REGA+ET High quality sequence stop: 422.
XX
FH   Key       Location/Qualifiers
FH
FT   source    1..438
FT             /organism="Mus musculus"
FT             /note="Vector: pT7T3D-Pac (Pharmacia) with a modified
FT             polylinker; Site_1: Not I; Site_2: Eco RI; 1st strand cDNA
FT             was primed with a Not I - oligo(dT) primer [5'
FT             TGTTACCAATCTGAAGTGGGAGCGGCCGCGGAAATTTTTTTTTTTTTTTTTTTTTTTTT
FT             T 3'], on equal amounts of mRNA from 2 13.5dpc and 2
FT             14.5dpc embryos [total RNA provided by Minoru Ko, Wayne
FT             State Univ., from 2 ]; double-stranded cDNA was ligated to
FT             Eco RI adaptors (Pharmacia), digested with Not I and cloned
FT             into the Not I and Eco RI sites of the modified pT7T3
FT             vector. Library went through one round of normalization,
FT             and was constructed by Bento Soares and M.Fatima Bonaldo."
FT             /sex="unknown"
FT             /strain="C57BL/6J"
FT             /clone="352061"
FT             /clone_lib="Soares mouse embryo NbME13.5 14.5"
FT             /dev_stage="13.5-14.5dpc total fetus"
FT             /lab_host="DH10B"
FT   mRNA      <1..>438
XX
SQ   Sequence 438 BP; 119 A; 117 C; 97 G; 105 T; 0 other;
     gcgaattcgc cttcagggcc cttactcact gctgcacttc tgcctcatca tatcctctgt        60
     gacgtgactc aaggtctacc tcatgctcac tctgcctgct tggatgagct gaagcgcacg       120
     tatgagttct tccggtactt tgaaactcag caccagtcag tccccagcg  cttatccaag       180
     actccgcaga agtcgaggga gctgagcaac gtccacacag cagtccgtag cttacagctc       240
     catctgaaag cactactgaa tgaagtatta attcttgaag atgaacttga aaagctcgtt       300
     tgtactaaag aaacacaaga gctgctatcg gaggcttatc cgatactaga gcagaaatta       360
     aaactgattg aaccacacgt tcaggccagc aacagttgct gggaggaagc catttctcaa       420
     gtggataaac tgctccgg
```

**Figure 2.3**    An example of the EMBL entry with the accession number W33426
(I.M.A.G.E. clone ID 352061).

## 2.2.1.9   Genome databases

Databases of the sequences of many complete or partial genomes are also
available. It is often not easy to find these sequences. The best place to start
looking is the WWW pages of the organization that sequenced the organism.
Some good places to find these organizations are lists and databases of useful
biological sites. Table 2.4 provides a few of these lists.

The genomes of many organisms can also be found at the EMBL and NCBI WWW sites: ftp://ftp.ebi.ac.uk/pub/databases/embl/genomes/ and ftp://ncbi.nlm.nih.gov/genbank/genomes/.

**Table 2.4.    Some genome resources.**

Genome Sequencing Projects
http://www-fp.mcs.anl.gov/~vgaasterland/genomes.html

Harvard Biological Laboratories
http://golgi.harvard.edu/

The BioCatalog
http://www.ebi.ac.uk/biocat/biocat.html

The HGMP Resource Centre
http://www.hgmp.mrc.ac.uk/genomeweb

The BioToolKit
http://www.biosupplynet.com/cfdocs/btk/btk.cfm

## 2.2.2    The main protein databases

SWISS-PROT is the prime protein sequence database. It is produced collaboratively by Amos Bairoch (University of Geneva) and the EBI. The data in SWISS-PROT are derived from translations of DNA sequences from the EMBL Nucleotide Sequence Database, extracted from the literature or submitted directly by researchers. It contains high-quality annotation, is nonredundant and is cross-referenced to many other databases, notably the EMBL nucleotide sequence database, PROSITE pattern database, the PDB protein structure database, the OMIM Mendelian Inheritance in Man database, the Mouse Genome Informatics (MGI) database, the restriction enzymes database REBASE, the G-protein-coupled receptor database GCRDb, and many species-specific genome databases and two-dimensional gel protein databases, among others.

TrEMBL (ftp://ftp.ebi.ac.uk/pub/databases/trembl/relnotes.doc) is a computer-annotated protein sequence database supplementing the SWISS-PROT database. It contains the translations of all coding sequences present in the EMBL database that are not yet integrated into SWISS-PROT. TrEMBL can be considered as a preliminary section of SWISS-PROT. It is split in two main sections: SP-TREMBL and REM-TREMBL.

SP-TREMBL (SWISS-PROT TrEMBL) contains the entries which should be eventually incorporated into SWISS-PROT. SWISS-PROT accession numbers have been assigned for all SP-TREMBL entries.

REM-TREMBL (REmaining TrEMBL) contains the entries that are not to be included in SWISS-PROT. REM-TREMBL entries have no accession numbers. It includes immunoglobulins and T-cell receptors, synthetic sequences, patent application sequences, small fragments and coding sequence translations where there is strong evidence to believe that the proteins are not real.

The general structure of an entry is identical in SWISS-PROT and TrEMBL.

## 2.2.2.1   How the data are entered

The entries in SWISS-PROT are derived from much the same sources as the nucleotide database entries, with the addition of translations of the coding sequences in EMBL entries.

Submissions to SWISS-PROT of directly sequenced peptides should be made via the site (http://www.ebi.ac.uk/submissions/submissions.html) at the EBI. Accession numbers will be provided for these sequences.

The EBI do not provide accession numbers, in advance, for protein sequences that are the result of translation of nucleic acid sequences. These translations will automatically be forwarded to SWISS-PROT from the EMBL nucleotide database and are assigned SWISS-PROT accession numbers on incorporation into TrEMBL.

## 2.2.2.2   Example entry

The general style and layout of the SWISS-PROT database entries is closely modelled on the EMBL nucleotide database format.

The accession number of the SWISS-PROT entry is unique and is not used in the nucleotide databases EMBL, GenBank or DDBJ. Differences include the following. The name of the entry is composed of two parts separated by an underscore character _. The first part is an abbreviation of the protein's family, function or name and the second part is an abbreviation of the species it is derived from, for example, '1433_CAEEL' is a 14-3-3-like protein from *Caenorhabditis elegans* and '1433_CHLRE' is a 14-3-3-like protein from *Chlamydomonas reinhardtii*. The 'GN' field holds the gene name. The 'DR' field is a cross reference to a related entry in another database. See Figure 2.4 for an example SWISS-PROT entry.

A major difference with EMBL entries lies in the feature table. The feature table is indicated by 'FT' at the start of the line. This is followed by the key name for that feature, then the start and end positions of the feature and then additional information on the feature. See Table 2.5 for a explanation of the SWISS-PROT feature keys.

```
ID   41BL_HUMAN     STANDARD;     PRT;     254 AA.
AC   P41273;
DT   01-FEB-1995 (REL. 31, CREATED)
DT   01-FEB-1995 (REL. 31, LAST SEQUENCE UPDATE)
DT   01-FEB-1996 (REL. 33, LAST ANNOTATION UPDATE)
DE   4-1BB LIGAND (4-1BBL).
OS   HOMO SAPIENS (HUMAN).
OC   EUKARYOTA; METAZOA; CHORDATA; VERTEBRATA; TETRAPODA; MAMMALIA;
OC   EUTHERIA; PRIMATES.
RN   [1]
RP   SEQUENCE FROM N.A.
RX   MEDLINE; 94374434.
RA   ALDERSON M.R., SMITH C.A., TOUGH T.W., DAVIS-SMITH T., ARMITAGE R.J.,
RA   FALK B., ROUX E., BAKER E., SUTHERLAND G.R., DIN W.S., GOODWIN R.G.;
RL   EUR. J. IMMUNOL. 24:2219-2227(1994).
CC   -!- FUNCTION: INDUCES THE PROLIFERATION OF ACTIVATED PERIPHERAL BLOOD
CC       T CELLS. MAY HAVE A ROLE IN ACTIVATION-INDUCED CELL DEATH (AICD).
CC       MAY PLAY A ROLE IN COGNATE INTERACTIONS BETWEEN T CELLS AND
CC       B CELLS/MACROPHAGES.
CC   -!- SUBUNIT: HOMOTRIMER (POTENTIAL).
CC   -!- SUBCELLULAR LOCATION: TYPE II MEMBRANE PROTEIN.
CC   -!- TISSUE SPECIFICITY: EXPRESSED IN BRAIN, PLACENTA, LUNG, SKELETAL
CC       MUSCLE AND KIDNEY.
CC   -!- SIMILARITY: BELONGS TO THE TUMOR NECROSIS FACTOR FAMILY.
DR   EMBL; U03398; G571323; -.
DR   PROSITE; PS00251; TNF_1; 1.
DR   PROSITE; PS50049; TNF_2; 1.
KW   CYTOKINE; TRANSMEMBRANE; GLYCOPROTEIN; SIGNAL-ANCHOR.
FT   DOMAIN       1     28     CYTOPLASMIC (POTENTIAL).
FT   TRANSMEM    29     49     SIGNAL-ANCHOR (TYPE-II MEMBRANE PROTEIN).
FT   DOMAIN      50    254     EXTRACELLULAR (POTENTIAL).
FT   DOMAIN      35     41     POLY-LEU.
SQ   SEQUENCE   254 AA;  26624 MW;  C68C1B27 CRC32;
     MEYASDASLD PEAPWPPAPR ARACRVLPWA LVAGLLLLLL LAAACAVFLA CPWAVSGARA
     SPGSAASPRL REGPELSPDD PAGLLDLRQG MFAQLVAQNV LLIDGPLSWY SDPGLAGVSL
     TGGLSYKEDT KELVVAKAGV YYVFFQLELR RVVAGEGSGS VSLALHLQPL RSAAGAAALA
     LTVDLPPASS EARNSAFGFQ GRLLHLSAGQ RLGVHLHTEA RARHAWQLTQ GATVLGLFRV
     TPEIPAGLPS PRSE
```

**Figure 2.4**   An example SWISS-PROT entry.

## 2.2.2.3   Other protein databases

GenPept (ftp://ncbi.nlm.nih.gov/genbank/genpept.fsa.Z) is GenBank's equivalent of TrEMBL. It is the automatic translation of all coding sequences present in the GenBank database.

OWL (http://www.biochem.ucl.ac.uk/bsm/dbbrowser/OWL/OWL.html) is a nonredundant protein sequence database produced from the following

**Table 2.5.**    The SWISS-PROT feature keys.

| Key | Description |
| --- | --- |
| CONFLICT | Different papers report differing sequences |
| VARIANT | Authors report that sequence variants exist |
| VARSPLIC | Description of sequence variants produced by alternative splicing |
| MUTAGEN | Site which has been experimentally altered |
| MOD_RES | Post-translational modification of a residue |
| LIPID | Covalent binding of a lipid moiety |
| DISULFID | Disulphide bond |
| THIOLEST | Thiolester bond |
| THIOETH | Thioether bond |
| CARBOHYD | Glycosylation site |
| METAL | Binding site for a metal ion |
| BINDING | Binding site for any chemical group |
| SIGNAL | Signal sequence (prepeptide) |
| PROPEP | Propeptide |
| CHAIN | Polypeptide chain in the mature protein |
| PEPTIDE | Active peptide released |
| DOMAIN | Domain of interest on the sequence |
| CA_BIND | Calcium-binding region |
| DNA_BIND | DNA-binding region |
| NP_BIND | Nucleotide phosphate binding region |
| TRANSMEM | Transmembrane region |
| ZN_FING | Zinc finger region |
| SIMILAR | Similarity with another protein sequence |
| REPEAT | Internal sequence repetition |
| HELIX | Secondary structure |
| STRAND | Secondary structure |
| TURN | Secondary structure |
| ACT_SITE | Amino acid(s) involved in the activity of an enzyme |
| SITE | Any other interesting site on the sequence |
| INIT_MET | The sequence is known to start with an initiator methionine |
| NON_TER | The residue at an extremity of the sequence is not the terminal residue |
| NON_CONS | Non consecutive residues |
| UNSURE | Uncertainties in the sequence |

source databases: SWISS-PROT, PIR, GenPept, NRL-3D (Bleasby *et al.*, 1994).

The International Protein Sequence Database (PIR) (http://www.mips. biochem.mpg.de/mips/pir-int_cd.html is a collaborative database from PIR, MIPS and JIPID that contains much the same sequence information as SWISS-PROT. However, it has a substantial amount of duplicated sequence entries, is hard to read and is not as well annotated. In particular, it lacks SWISS-PROT's superb cross-referencing to other databases.

NRL_3D (http://www.rcsb.org/pdbl) is produced by PIR from sequence and annotation information extracted from the Brookhaven Protein Databank (PDB) of crystallographic 3D protein structures. It is useful to do similarity searches against this database, because a match between your sequence and a NRL_3D entry shows that there is a similar protein sequence to your sequence whose 3D structure is known. This is useful in any further work to predict the 3D structure of your sequence.

The Kabat Database of Sequences of Proteins of Immunological Interest (http://immuno.bme.nwu.edu/) is a database of sequences involved in the immune system. It contains groupings of related amino acid and nucleotide sequences of:

- signal sequences of light chains, heavy chains, T-lymphocyte receptors, related proteins
- variable region of light chains, heavy chains, T-lymphocyte receptors for antigen
- constant region
- major histocompatibility antigens Class I
- I region gene products Class II
- related proteins
- D and J minigenes
- Pseudogenes.

## 2.3   RATE OF DATABASE GROWTH

The databases grow exponentially. The SWISS-PROT database doubles in size about every 40 months and the nucleotide databases double in size about every 14 months (Fig. 2.5). There has been a marked increase in the growth rate of the nucleotide database since 1994 when EST sequences began to be added in significant numbers.

This remarkable growth rate will probably continue for as long as sequencing technology improves and funding for the genome projects continues. The completion of sequencing the human and mouse genomes may signal the beginning of the end of the exponential growth phase for the nucleotide databases. The nucleotide databases size might then be between 50 gigabytes and 200 gigabytes, depending on how long the genomes take to complete.

The ever-increasing requirements for disk-space and computer power have kept pace remarkably well over the years with the ever-dropping price of computer processing power and hardware. However, the number and complexity of the searches and analyses that researchers need to do has also increased, so

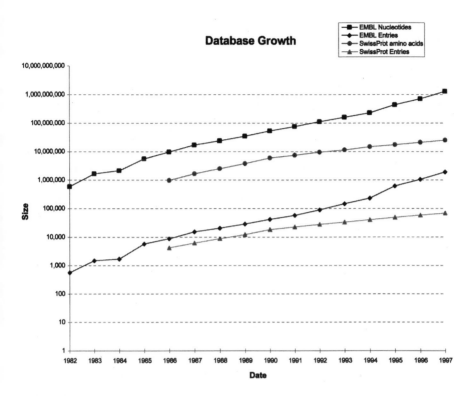

**Figure 2.5**   Database growth.

the amount of computer power required by the average organization has steadily increased.

This has lead to problems for some smaller organizations, leading them to stop holding the databases locally or to stop updating them frequently. There is therefore an increasing trend to rely on a few major sites which make the databases available to researchers in various ways over the Internet.

## 2.4   Problems with the data

The nucleotide database sequences contain a significant amount of contamination by vector, bacterial, rRNA and other spurious sequences. For example, a simple BLAST search through some of the divisions for similarity to the BlueScribe KS Minus cloning vector sequence finds 103 excellent matches in the Human division, 77 matches to the Rodent division and well over 1000 matches in the EST division. The database organizations work hard to eliminate such contamination by informing the authors that it has been detected, but it can still be found; sometimes, the authors do not care enough to correct their sequences.

The error rates of nucleotide sequences that are completed by major sequencing organizations is very good (1/10 000) and improving, apart from the 1% error rate of the single-read sequences of the EST and HTG divisions. However, many protein sequences have been derived by translating DNA sequence. Sometimes frame-shift errors in translation leads to a completely erroneous protein sequence. About 5–10% of sequences in SWISS-PROT have been found to have frame-shift errors.

When looking at matches to protein sequences, remember that the vast majority of these have been produced by translation of the corresponding nucleic acid sequence and that the prediction of exons is error-prone – it is only about 85–90% accurate at best.

About 30% of these protein database entries may therefore have a missing exon or an intron erroneously translated and inserted in their sequence. If you try to find exons in a genomic nucleic acid sequence by doing a similarity search against the protein database and by using one of the many available exon prediction programs, you will often find conflicts between the exonic regions predicted by the two methods. However, you have no justification in assuming that the prediction programs are wrong and that the protein match shows the correct pattern of exons. Equally, you have no justification in assuming the converse. (Do the laboratory work to confirm any prediction!)

It is common to assign function to a sequence by finding potentially homologous sequence entries. Thus, if the database entry A shows similarity to sequence B, the annotation of B may be based on the function given to A in its annotation. This is very commonly done with ESTs and other large-scale sequencing projects. The problems occur when the annotation of A is updated to reflect further work done on the sequence. The annotation of B will not be updated and its annotation will continue to be erroneous.

Other problems with annotation occur when 'third-hand' or 'fourth-hand' assignments of function (sequences C, D, etc.) are all based on the annotation of a first sequence (A) to which the later sequences in the chain have no discernible sequence similarity, and possibly no homology, even though each link in the chain may show similarity (A–B, B–C, C–D, etc.).

The annotation of features of sequences has often been done several years ago and not been updated since. The biological techniques and computer software to find features in sequences are continually improving and so many feature tables may now be incomplete or inaccurate.

Most sequences in the databases are what they claim to be, but be aware that there are many small and some large errors in the sequences and annotation. Do not blindly accept database search results. If you get anomalous results, stop and think for a while.

## 2.5   ACCESSING THE DATABASES

It is possible that your site may have set up one of the many available packages for analysing and searching gene sequences. If so, there may be considerable expertise on-site in the use of these packages and you should make use of this.

You should investigate the following publicly available WWW sites for keyword and similarity searches:

(1) Entrez (http://www3.ncbi.nlm.nih.gov/Entrez/) provides good cross linking between the nucleic, and protein databases and with the Medline bibliographic database. A very powerful feature is the ability to find other entries like the one you have already found.

(2) SRS, the Sequence Retrieval System (http://srs.ebi.ac.uk) provides a powerful means of finding entries in related sets of databases.

(3) BLAST (http://www.ncbi.nlm.nih.gov/BLAST/) and FASTA (http://www2.ebi.ac.uk/fasta3/) are publicly available sites providing access to these popular sequence similarity searching programs.

## REFERENCES

Bairoch A. & Apweiler R. (1997) *Nucl. Acids Res.* **25**, 31–36.
Benson D.A., Boguski M.S., Lipman D.J., Ostell J. & Ouellette B.F.F. (1998) GenBank. *Nucl. Acids Res.* **26**, 1–7.
Bleasby, A.J., Akrigg, D. & Attwood, T.K. (1994) *Nucl. Acids Res.* **22**, 3574–3577.
Gonzalez I.L. & Sylvester J.E. (1997) *Gen. Res.* **7**, 65–70.
Lennon, G., Auffray, C., Polymeropoulos M. & Soares M.B. (1996) *Genomics* **33**(15), 1–152.
Ouellette B.F.F. & Boguski M.S. (1997) *Gen. Res.* **7**, 952–955.
Stoesser G., Moseley M.A., Sleep J., McGowran M., Garcia-Pastor M. & Sterk P. (1998) *Nucl. Acids Res.* **26**, 8–15.

# 3 Phenotype, Mutation and Genetic Linkage Databases and Their Links to Sequence Databases

*Richard Cotton*

Mutation Research Centre, St Vincent's Hospital, Fitzroy 3065, Melbourne, Australia

## 3.1 INTRODUCTION

The importance of mutation databases in human research has been grasped only relatively recently. Thus, the reader will gather the impression that the area is in a state of flux at the present time, while guidelines and rules are being established. This chapter will focus on the human mutation databases (owing to the author's expertise), but all living organisms have been subject to specific variation that has been recorded over the centuries. Large and systematic listings of variation at the phenotype level and later at the genotype (mutation) level began in the 1950s with the isolation of auxotropic and other variants of the bacterium *Escherichia coli*. More recently, this has extended to possibly hundreds of organisms, such as the nematode *Caenorhabditis elegans*, the plant *Arabidopsis thaliana* and more notably HIV and the mouse. Naturally, there are potentially as many phenotypes and mutation databases, as there are organisms. For the reader interested in organisms other than human, a search on the World Wide Web (WWW) should reveal databases of interest. For example, mouse and *Arabidopsis* variation can be found in the The Mouse Genome Database (MGD, http://www.informatics.jax.org) and the *A. thaliana* Database (AtDB, http://genome-www.stanford.edu/Arabidopsis/) respectively.

*Genetics Databases*
ISBN 0-12-101625-0

## 3.2    LEVELS OF VARIATION

Examination of organisms from previous centuries to the present day has led to their classification. Division into the orders, families, genus and species are well established, although the boundaries are sometimes blurred and debatable. It is assumed there are numerous 'genetic' differences between species, but there may be an order of magnitude less difference between individuals in species. This level of difference is then the focus of mutation databases of humans, where in the approximately 50 000 genes there may be tens of thousands of differences in the genes between normal individuals by way of single base-pair substitutions. This type of variation is usually harmless and is commonly referred to as a polymorphism (see below), but nevertheless, may cause, for example, differences between individual's eye colour.

Disease-causing base changes are much rarer between individuals and often are referred to as mutations (see below).

Lists of inherited diseases in humans started to be collected by Victor McKusick in the 1960s. This was essentially a phenotype database, which named syndromes and gave them a number. Separately, a mutation causing human disease was defined in 1949 in the globin protein and by 1976, 200 such disease-causing mutations had been listed. These were all different but caused the same disease and patients carrying them had some differences in the disease they displayed, despite having a mutation in the same gene. This then is yet another layer of variation. While not the topic of this chapter, it is possible to have even another layer of variation as siblings with the same mutations and same disease, can have radically different phenotype due to what are referred to as modifiers, or modifying genes (Summers, 1996).

The main topic of this chapter is databases of the sequence differences found between human individuals in specific genes. If there are 50 000 genes, each with at least 100 variations causing disease, this means that all the databases might together contain five million mutations causing disease. This excludes normal variants, or polymorphisms.

## 3.3    DEFINITIONS

- *Phenotype*. In the context of human disease, this usually refers to the normal or the appearance of individuals with disease caused by base changes in the gene that causes the disease, e.g. mutant, or normal phenotype.

- *Genotype*. In the context of human disease, this usually refers to the status of the disease genes in an individual. For example, in a recessive disease such as phenylketonuria (PKU), two copies of a defective gene are necessary to cause disease. The genotype of the affected individuals is said to be homozygous mutant.

■ *Base change.* This refers to any change in the genomic DNA between individuals, which may be a substitution, deletion, or insertion that may, or may not, cause disease.

■ *Polymorphism.* This term in common usage in human genetics refers to a base change, which does not affect phenotype when referring to a disease gene. A formal definition of this term defines it as a change, which is present in the normal population at greater than 1%.

■ *Mutation.* The strict definition of this term is any base change and is used throughout biology for this event. However, it is confusing that in common usage in human genetics and clinical genetics; this term is used for those base changes causing disease.

■ *Allelic variant.* This term has been suggested by McKusick to supplement 'mutation' and 'polymorphism' and thus, avoid confusion. One of the main reasons for use of this term is that when a base change is described in a gene, it is often not immediately obvious whether it is deleterious or not. It may be some time before it is proven disease causing, i.e. a mutation or harmless, i.e. a polymorphism.

■ *Polygenic disorder.* Sometimes referred to as common diseases that include heart disease, asthma, etc., these diseases are caused by base changes at several loci.

Previously, polymorphisms in genes have been regarded as unimportant in human genetics, as they do not cause disease. (They have been important in linkage studies, but most polymorphisms used for this purpose have been anonymous and extragenic.) However, their importance has recently been realized and they are being actively sought after and recorded (see below).

## 3.4   TYPES OF DATABASES

Theoretically, phenotype databases record phenotype differences and mutation databases record genotype differences. Before genes for the various genetic disorders were cloned and before mutational changes were identified, inherited disease databases were necessarily only phenotypic. However, more recently, as genes have been cloned and mutations identified, these databases have become both phenotypic, or genotypic. Before the days of the computer and the World Wide Web (WWW), all databases were kept on cards or in books. These then graduated to computers on desks, and then finally to the WWW. Not all databases are on the WWW, but there is a strong move in this direction.

### 3.4.1    General or central databases: collecting genotypes and phenotypes in all inherited disease

#### 3.4.1.1    The Metabolic and Molecular Basis of Inherited Disease (Scriver et al., 1995)

This book, which is in its 7th edition was first published in 1960, documents the phenotype, i.e. metabolism in inherited disease, and later added genes and mutations. This work is in three large volumes and, thankfully, it is also on disk but not yet on the WWW. In addition to many general chapters and listings, each disease has the phenotype described in detail and has listings of disease-causing mutations in genes, as well as presumed harmless polymorphisms for some genes. It is necessarily updated in each new edition that appears about 4 years apart.

#### 3.4.1.2    On-line Mendelian Inheritance in Man (OMIM: http://www3.ncbi.nlm.nih.gov/omim/)

Victor McKusick conceived this database in 1960 and first published a listing of X-linked traits in 1962. Since that time, a printed version under the title *Mendelian Inheritance in Man* has appeared. Subsequently, this was placed on the WWW and is known as 'On-line Mendelian Inheritance in Man' (OMIM). Each disease entity has a number (MIM number) and there is a brief description of the phenotype. Mutations are also listed, but only in limited number, concentrating mainly on those which make a specific point; the listing claims to be representative not exhaustive. Key recent reviews and references are given. There are no direct links from each entry, but there are links at the beginning to useful general resources and to locus-specific databases (see 3.4.2).

#### 3.4.1.3    Human Gene Mutation Database (HGMD: http://www.cf.ac.uk/uwcm/mg/hgmd0.html)

This database began in the printed word in the 1980s, as the work of David Cooper and colleagues. This was a listing of mutations and their immediate context, the authors' aim being to study the frequency and other characteristics of particular mutational events. The entries were each provided with a reference. It is in this form today and as such is an index of all *published* mutations, which is an extremely useful resource for those discovering mutations in the laboratory and in the clinic. Mutation maps and a listing of locus-specific databases with links are also provided. Each gene entry has links to the OMIM entry, GDB (see below) entry and locus-specific databases (see 3.4.2).

### 3.4.1.4  Genome Database (GDB: http://gdbwww.gdb.org/)

This database began to list markers and maps of the human genome and homologies with other mammalian genomes. More recently, it has been recording variation in genes, but the effort was not comprehensive. Recently, a collaboration between HGMD was announced, so that the listing in GDB could be kept up to date.

### 3.4.1.5  Genatlas (http://www.citi2.fr/GENATLAS/welcome.html)

Started by Jean Frezal in Paris, France, this database concentrates more on phenotypes. It contains information on over 7800 genes, 18 100 markers and 1800 phenotypes. Some mapping data is given. Homologies, motifs, etc., are given. Chromosomal rearrangements or break-points associated with developmental abnormalities are given. Mitochondrial disorders are also given, but mention of other mutations is made only in the references and components of polygenic traits. Much linkage data is given around loci and there are links to locus-specific databases, GDB and comparative maps.

### 3.4.1.6  Location Data Base (LDB: http://cedar.genetics.soton.ac.uk/public_html/)

The LDB is a database for constructing fully integrated genetic and physical maps of the human genome (Collins *et al.*, 1996).

### 3.4.1.7  On-line Mendelian Inheritance in Animals (OMIA: http://www.angis.su.oz.au/Databases/BIRX/omia/)

This database sets out to be for the animal world what OMIM is for human disorders. There are 114 species listed in the database and there are genetic maps for 14.

There are nine headings under disorders or traits, such as those with a molecular basis, with linked marker, known gene or peptide, useful as animal models, inborn errors of metabolism, etc.

### 3.4.1.8    EMBL/EBI Mutation Projects (http://www2.ebi.ac.uk/mutations/)

Michael Ashburner has recently initiated two mutation programs to provide data and assistance to the mutation community. This program provides resources, such as links to relevant mutation databases and runs a server with an SRS (Sequence Retrieval System) program, so that individual databases loaded on it can be searched as one (see Web site). It also has a total listing of all mutations in featured databases under the heading EMBL CHANGE that listed 32 863 entries on 15 January, 1999. Central mutation database software is currently being designed.

### 3.4.1.9    SwissProt (http://www.ebi.ac.uk/ebi_docs/swissprot_db/swisshome.html)

SwissProt is a database of protein sequences, but many mutant sequences have been described and thus this database contains mutations.

## 3.4.2    Locus-specific databases

In its simplest form, a locus-specific database has a listing of mutations in a specific gene. The base changes listed may be either harmful, harmless, or something in between or status unknown.

The information in these databases includes disease name, gene name, OMIM number, etc. The core information is the formal name based on the primary sequence change and the trivial name based on the amino acid change in the case of relatively simple mutations. Much additional information is available, such as phenotype of the patient, geographic association, frequency and the mutation in the other allele if relevant.

It is unknown how many such databases exist. They may range from printed word, e.g. the globin mutations of Huisman *et al.* (1996), which is a 420 page printed book, to listings in a scientist's computer, to Web sites run by individuals or consortia. Many journals have reviews of mutations in genes and in databases and these often occur as a prelude to construction of a Web site. The journals *Nucleic Acid Research* and *Human Mutation* feature such reviews. The EBI Web site lists and has links to, and has on their server 41 such databases. On 16 June, 1999, the HGMD database had links to 59 and the Mutation database initiative (see below) had a listing of at least 96 available on the web and also had links. (Some databases contain information on numerous genes, e.g. the tetrahydrobiopterin deficiency database and some genes have more than one database, e.g. *p53*.)

Examination of the numerous databases indicates that the curator ranges from an individual on a shoestring budget to someone representing a large consortium or with a large budget. The following is a discussion of six key databases.

## 3.4.2.1   Haemoglobin (http://globin.cse.psu.edu/)

The variants listed by Huisman *et al.* (1996) can be seen in Table 3.1. A total of 693 variants of the globin genes had been documented in 1996. In 1976 it was a mere 200. Because of the structural interest in the field, some data on the structural effect is given. Other data includes patient phenotype, laboratory phenotype-defining studies on electrophoresis, chromatography, peptide studies, oxygen studies and stability. Often, DNA studies are not given as the mutation may have been defined at the protein level. Ethnic origin is given, as well as quantity of mutant molecules in the heterozygote.

A recent agreement to place the data on the globin web site will assist greatly in the analysis and search of the data on diseases of the four globin genes.

**Table 3.1.**   The number of Hb variants listed in the syllabus of human haemoglobin variants (January 1996).

| Type | Number |
|------|--------|
| $\alpha$ Chain variants | 199 |
| $\beta$ Chain variants | 335 |
| $\gamma$ Chain variants | 68 |
| ($^{G}\gamma$ = 38; $^{A}\gamma$ = 20; unknown = 3; special = 7) | |
| $\delta$ Chain variants | 28 |
| Variants with two amino acid replacements | 18 |
| ($\alpha$ = 1; $\beta$ = 17) | |
| Variants with hybrid chains | 10 |
| Variants with elongated chains | 13 |
| (at the *C*-terminus = 9; at the *N*=terminus = 4) | |
| Variants with deletions (15); with insertions (4); | 22 |
| with deletions and insertions (3) | |
| Total | 693 |

### 3.4.2.2    Cystic Fibrosis Mutation Database
### (http://www.genet.sickkids.on.ca/cftr/)

This locus specific database run by L-C. Tsui for a large consortium contains at its core:

(1)    Lists of mutations which cause cystic fibrosis (CF), which in June 1999 totalled 870;
(2)    Lists of coding region 'polymorphisms' (68) and;
(3)    Noncoding region polymorphisms (55). No phenotype information is given, which is unusual in a locus-specific database.

Other information includes cDNA and gene sequence, population variation and newsletters and the site has links to other CF databases and information.

### 3.4.2.3    Phenylketonuria Database
### (http://www.mcgill.ca/pahdb/)

Phenylketonuria (PKU) has been a model disease among all inherited diseases for various aspects such as detection, treatment, etc. and it now serves as a model locus-specific database under the direction of Charles Scriver and colleagues for a consortium which contained 91 members in 1999.

The tables list the 384 mutations (current at June 1999) and independent discoveries of each, as well as haplotypes, haplotype association, frequency in various populations, *in vitro* expression analysis and genotype phenotype analysis.

There is a map of the mutations and two items for patients, a PKU resource booklet for patients and families and a clinical essay for patients. Details of the mouse model of PKU and more recently structural information is also present.

This database perhaps contains the largest and most complete set of data on mutations causing inherited disease. Recently, the Web site has changed from tabular format to relational format.

### 3.4.2.4    Haemophilia B
### (http://www.umds.ac.uk/molgen/haemBdatabase.htm)

This database has 1918 patient entries in the latest version and lists the patients under code names and then describes the amino acid and base change with a phenotype, which, in this case, are clotting value and antigen level. A reference is given together with a comment. For the purpose of gathering mutations, the world is divided into nine regions with particular individuals being responsible for them.

## 3.4.2.5 TP53

   (1)   http://perso.curie.fr/tsoussi/p53_database.html
   (2)   http://www.iarc.fr/p53/Homepage.htm
   (3)   http://sunsite.unc.edu/dnam/mainpage.html

Because *p53* mutations are usually somatic, causing malignancy in almost all types of cancer, it is a special case. It is also unusual in that three different databases exist. The IARC database (No. 2) contains germ-line mutations, polymorphisms and tumour mutations. The Beroud and Soussi (No. 1) database contains polymorphisms, phylogeny and details of mutation analysis, as well as the main database of mutations. This database contains various analyses of the data. Because of the epidemiological value of the study of *p53* mutations, multiple entries for each mutation are kept.

   These two databases contain over 8000 and 4200 records respectively. Another database (http://www.mayo.edu/research/papers/P53%20Mutations) is available but contains both TP53 and APC mutations in the one table. There are 2336 entries.

## 3.4.2.6 Mitochondrial Mutation Database (http://www.gen.emory.edu/mitomap.html)

This database is a special case because the mitochondrial genome has many genes and possibly each one could have a mutation database of its own.

   The site contains almost the complete molecular biology and genetics of the mitochondrial genome. Polymorphism sites, continent specific variants, missense mutations, mRNA, tRNA mutations, deletions, insertions and complex rearrangements are covered. In short, it is almost a text book on mitochondrial biology and genetics.

## 3.4.3 Linkage databases

Where linkage is the only route to identify cloning a disease gene, linkage maps are an essential part of the process (the more so for the immense task of mapping and sequencing the whole genome). This linkage can be physical or can be derived from genetic studies, which can give an estimate of the proximity of markers along the chromosomes.

## 3.4.3.1 Généthon (http://www.genethon.fr)

Among other things, this gives the current map of the human genome.

### 3.4.3.2    Co-operative Human Linkage Centre
### (http://www.chlc.org/)

This mainly exists to develop statistically rigorous high-quality maps that have a high proportion of easy to use microsatellite markers.

### 3.4.3.3    CEPH Genotype Database
### (http://www.cephb.fr/cephdb)

CEPH maintains a special collection of families that are used as a resource by scientists around the world for linkage studies of new or old markers. A database is maintained of these markers and on December, 1998 contained data on 11 995 markers assigned to all known chromosomes.

## 3.4.4    Genetic resources databases

Once a disease gene has been sequenced the main activities, which can begin are:

(1)    The accumulation of mutations; and
(2)    The diagnosis of patients.

These activities require expertise in mutation detection access to mutation databases, access to companies performing tests and resources for patients.

### 3.4.4.1    Alliance of Genetic Support Groups
### (http://medhelp-org/www/agsg.hgm)

This site contains a list of around 300 support groups that patients and clinicians may consult.

### 3.4.4.2    Other databases for support for those with genetic disorders

The following is a list of other resources of those with inherited disorders:

- March of Dimes: http://modimes.org/
- *Canadian Genetic Diseases Network:* http://www.cgdn.generes.ca
- *Council of Regional Networks (CORN) for Genetic Services:* http://www.cc.emory.edu/PEDIATRICS/corn/corn.htm

- *Genetics Syndrome Support Group listings:* http://members.aol.com/dnacutter/sgroup.htm
- *Rare diseases in children:* http://mcrc4.med.nyu.edu/2murphp01/homenew.htm
- *National Organization of Rare Disorders:* http://www.NORD_RDB.com/2orphan

### 3.4.4.3   Polymorphism Database

Polymorphisms have been important in human genetics since DNA studies began. They have been used for linkage to disease genes in diagnostic studies, in gene discovery and, most recently, in whole-genome scans. The possibility that intragenic, including expressed polymorphisms can be used to discover components of polygenic disorders or modifier genes has recently been realized. This has led to formation of databases especially for polymorphisms. There are now four available. NCBI initiated dbSNP (http://www.ncbi.nlm.gov/SNP/). HGBASE (http://hgbase.interactiva.de/) is a joint venture between Uppsala University of Medical Genetics and Interactive Biotechnologie. The Human SNP database (http://www.genome.wi.mit.edu/SNP/human/) is maintained at The Whitehead Institute/MITCentre for Genome Research. The SNP database (http://www.ibc.wustl.edu/SNP/) is maintained at Washington university, St. Louis.

## 3.5   USING MUTATION DATABASES

There are many types of workers who would use databases in many different ways and some examples are given.

The diagnostic laboratory when detecting a mutation for the first time will wish to know if it has been found before and if it has been proven to cause disease when found in that gene, so that the hunt for mutations might be terminated. The clinician may further wish to read the papers where the mutation has been described, so that prognostic information can be gleaned for the patient.

The diagnostic laboratory will first try HGMD to see if the mutation is published and if not, will then move to the locus specific database, if one exists, to see if it has been described but not published. The clinician if dealing with a rare disease may first visit OMIM or use a copy of MIM to read about the disease and if the mutation is not in the listing in the section go to HMDB and then the locus-specific database. If the mutation is listed as unpublished the investigator may have to contact the locus-specific database curator to obtain a contact address for those describing the mutation.

If the mutation is not described, the investigators will then need to publish it, either via the journals or via the locus-specific databases as an entry.

Those wishing to set up diagnostic tests for a number of mutations described in genes will wish to find which have been described and which are the most frequent in a particular ethnic group under consideration. This type of information is only likely to be contained in locus-specific databases, but not all of them have this data and, like HGMD and OMIM, only describe the first description of that mutation in the literature.

Biochemists interested in the protein product from the gene are likely to look at the databases to get an indication of which amino acids might be important in the function of the protein.

Patients concerned about an inherited disease in their family may wish to read more and track down, as described above, the clinician looking after patients with the same mutations as them, or read patient-related material that might be in the locus-specific database. They may also wish to contact the support groups.

## 3.6    EXERCISES

The following exercises are designed to not only familiarize the reader with how to use the databases, but also to indicate the nonuniformity of the databases and the deficiencies of the system as a whole.

### 3.6.1    The number of mutations in a gene

Count the number of mutations in the PKU gene and the CF gene in OMIM, HGMD and the locus-specific database. Why do differences exist and in which database is it easier to count them?

### 3.6.2    Clinical information for patients

Visit 10-20 locus-specific databases through links at OMIM or HGMD to find how many provide up-to-date information for patients.

### 3.6.3    Entering mutations

When accessing the databases in Point 6.2, find which ones provide forms for data entry.

## 3.6.4　Number of times a mutation is described

Visit CF, PKU and BTK base (http://helsinki.fi/science/signal/btkbase.html) to determine which databases record independent occurrences of mutations.

## 3.7　THE HUGO MUTATION DATABASE INITIATIVE

The ideal human mutation database would contain up-to-date lists of base changes found in genes whether published or unpublished. These should be easily accessible by the Web and the data should be accurate. Published data should be accurate, but unpublished data may not be guaranteed. Collecting this information is time-consuming and expensive and unlikely to be performed by a central organization if there prove to be millions of base changes to be recorded. An initiative began, in 1994, to assist in the process of recording and distributing this information. This is now known as the HUGO Mutation Database Initiative and has the assistance of the March of Dimes. This group has a membership of approximately 550 and exists to promote locus-specific databases, and central databases, working together to create a system most useful for the community. The group has considered and is considering such vital topics as nomenclature of mutations, quality control, copyright, content and software of databases, etc. History and progress of the Initiative can be seen on its Web site: http://ariel.ucs.unimelb.edu.au:80/~cotton/mdi/.htm.

Readers wishing to contribute to this Initiative can join the association via form on the Web site.

## 3.8　CONCLUSION

It is an enormous but vital task to list all base changes in the human genome. Possibly, in a hundred years it will be complete. However, efforts are underway between scientists, journals, central databases and locus-specific databases to begin this task. They are assisted by companies and grant-giving bodies and this assistance needs to continue for some time.

## ACKNOWLEDGEMENTS

Valerie Mehl and Rania Horaitis are thanked for their help in the preparation of the manuscript.

# REFERENCES

Collins A., Frezal J., Teague J. & Morton N.E. (1996) *Proc. Natl. Acad. Sci. USA* 93, 14771–14775.

Huisman T.H.J., Carver M.F.H. & Efremov G.D. (1996) *A Syllabus of Human Hemoglobin Variants*. The Sickle Cell Anemia Foundation, Augusta.

Scriver C.R. *et al.* (1995) *The Metabolic and Molecular Basis of Inherited Disease*. McGraw-Hill, New York.

Summers K.M. (1996) *Hum. Mut.* 7, 283–293.

# 4 DNA Composition, Codon Usage and Exon Prediction

## Roderic Guigó

Informàtica Mèdica, Institut Municipal d'Investigació Mèdica and Departament d'Estadística, Universitat de Barcelona, Barcelona, Spain

## 4.1 INTRODUCTION

In this chapter, we review the sequence based measures indicative of protein-coding function in genomic DNA. As the genome projects enter the large-scale sequencing phase, computer programs become essential to identify protein coding genes in large uncharacterized genomic sequences, typically of tens of thousands or even hundreds of thousands of nucleotides, with efficiency and reliability. At the core of all gene identification programs there exist one or more coding measures. Such programs rely mostly on additional information, such as potential sequence signals involved in gene specification and sequence similarity database searches, and use very complex frameworks for its integration. However, a good knowledge of the core coding statistics is important in order to understand how gene identification programs work, and to interpret their predictions. Here, we will review a few of the most important coding measures and illustrate through examples the details involved in their computation.

A coding statistic can be defined as a function that computes given a DNA sequence a real number related to the likelihood that the sequence is coding for a protein. Since the early eighties, a great number of coding statistics have been published in the literature. Most such coding statistics measure either codon usage bias, base compositional bias between codon positions, or periodicity in base occurrence (or a mixture of them all). Exhaustive reviews can be found elsewhere (see, for example, Gelfand, 1995, and the references therein). Here, we follow loosely the critical review by Fickett & Tung (1992). Our classification of coding measures is, however, slightly different. The main distinction here is

*Genetics Databases*
ISBN 0-12-101625-0

between measures dependent on a model of coding DNA and measures independent of such a model. The model of coding DNA is always probabilistic, allowing one to compute the probability of a DNA sequence, given that the sequence is a coding sequence. Although in practice the values (scores) of a given coding statistic in a query sequence can be computed in a number of different ways, for the model-based coding statistics we will compute scores based on such a probability. Indeed, given a query sequence, we will compute the probability of the sequence using the model of coding DNA and using an alternative model or noncoding DNA (which, for illustration purposes, will be simply random DNA). We will take the logarithm of the ratio of these two probabilities, the log-likelihood ratio, as the score of the coding statistic in the query sequence.

Model-dependent coding statistics are likely to capture more of the specific features of coding DNA since they are dependent on more parameters. Therefore, model-dependent coding statistics may be more powerful in discriminating coding from noncoding DNA. However, model dependent coding statistics, require of a representative sample of coding DNA from the species under consideration in order to estimate the parameters of the model. The more complex the model the more sensitive it is to sample bias and size. Conversely, model-independent coding statistics capture only the 'universal' features of coding DNA. Since they do not require a sample of coding DNA, they can be used even in absence of previously known coding regions from the species under consideration.

To illustrate the coding statistics reviewed here, we will use a few test sequences. As a genomic test sequence, we will use a 2000 bp DNA sequence from the human genome coding for the β-globin gene. The sequence has been extracted from the GenBank entry HUMHBB (EMBL entry HSHBB), from positions 62 001 to 64 000. The human β-globin gene has three coding exons in positions 187–278, 409–631, and 1482–1610, relative to the 2000-bp test sequence. We have also extracted two subsequences from this test sequence. The complete sequence of the second coding exon of the β-globin gene, which is 223 bp long, and a 223 bp long sequence from the middle of the second intron (from positions 800–1022 of the test sequence). These two sequences will serve as exonic and intronic test sequences. We have also extracted from GenBank Release 93 (February 1996), a set of 450 human genes following the protocol described in Burset & Guigó (1996; see also Guigó (1997b)). From this set, exons longer than 100 bp and introns longer than 100 bp and shorter than 2500 bp were retained. This resulted in 1753 introns and 1761 exons. The distribution of the scores of the different coding statistics analysed here will be plotted in these two sets of sequences. Results obtained in the exonic, intronic and genomic test sequences are only for purposes of illustration, and differences in the performance of different coding statistics can not be inferred from them. Readers interested in a rigorous comparative benchmarking should refer to Fickett & Tung (1992).

We will try to avoid complex mathematical formulas to describe the algorithms, but we will indicate how to calculate them, so that interested readers can reproduce the computations. Although, we will thus keep mathematical formalisms to a minimum, it will be useful to maintain a consistent notation through the chapter. Thus,

$$S = S_1 S_2 \ldots S_l$$

will denote a DNA sequence of length $l$, while $S_i$ ($i = 1 \ldots l$) will denote the individual nucleotides. For instance if

$$S = \text{AGGACGGGATCA}$$

then

$$S_1 = \text{A}, \quad S_2 = \text{G}, \quad \ldots \quad S_l = \text{A}, \quad \text{and } l = 12$$

A DNA sequence can be partitioned in a sequence of consecutive nonoverlapping codons in three different ways depending on the nucleotide in the sequence on which the grouping of nucleotides into codons starts (that is, the sequence can be read in three different frames). If $C$ is a sequence of codons

$$C = C_1 C_2 \ldots C_m$$

$C_j$ will denote the codon occupying position $j$ in the sequence. If $S$ is a nucleotide sequence, we will use $C_s^i$ (or simply $C^i$, $i = 1, 2, 3$) to denote the sequence of codons that results when the grouping of nucleotides from $S$ into codons starts at nucleotide $i$. We will use $C_j^i$ to denote the codon occupying position $i$ in the decomposition $j$ of the sequence. For example, if $S$ is the sequence above, then

$$
\begin{aligned}
C_1^1 &= \text{AGG} & C_2^1 &= \text{ACG} & C_3^1 &= \text{GGA} & C_4^1 &= \text{TCA} \\
C_1^2 &= \text{GGA} & C_2^2 &= \text{CGG} & C_3^2 &= \text{GAT} \\
C_1^3 &= \text{GAC} & C_2^3 &= \text{GGG} & C_3^3 &= \text{ATC}
\end{aligned}
$$

Alternatively, if $c$ is a codon, we will use $c[k]$ to denote the nucleotide occupying position $k$ in the codon. For example for the above

$$C_1^1[1] = \text{A} \quad C_2^2[3] = \text{G} \quad C_1^3[2] = \text{A}$$

## 4.2 MEASURES DEPENDENT ON A MODEL OF CODING DNA

All the measures dependent on a probabilistic model of coding DNA can be computed in a uniform way through the computation of the probability of the sequence given the model. That is, given a probabilistic model of what coding DNA is, the codon usage table, for example, we can compute the probability of a sequence of nucleotides $S$, assuming the sequence is coding in a given frame. We will use $P_i(S)$ to denote the probability of the sequence of nucleotides $S$, given that $S$ is coding in frame $i$ ($i$ = 1, 2, 3). Alternatively, we can compute the probability of $S$ given a model of noncoding DNA. We will use $P_0(S)$ to denote such a probability. $P^i(S)/P_0(S)$ is a likelihood ratio: the ratio of the probability of finding the sequence of nucleotides $S$, if $S$ is coding in frame $i$ over the probability of finding the sequence of nucleotides $S$, if $S$ is non-coding. To measure the coding potential of sequence $S$ in frame $i$ given the model of coding DNA, we will compute the natural logarithm of this ratio, the log-likelihood ratio,

$$LP^i(S) = \log \frac{P^i(S)}{P_0(S)}$$

If $LP^i(S) > 0$, then the probability of the sequence of nucleotides $S$ is higher assuming that $S$ is coding in frame $i$, than assuming that $S$ is noncoding in frame $i$, while if $LP^i(S) < 0$, then the probability of $S$ is higher assuming that $S$ does not code in frame $i$ than assuming that $S$ is coding in frame $i$. Given a sequence problem, $S$, we compute the log-likelihood ratios for $S$ in the three frames. If the sequence is coding, the log-likelihood ratio will be larger for one of the frames than for the other two.

Throughout the chapter, we will assume noncoding DNA to be simply random DNA with nucleotide equiprobability and independence between positions. It could be argued that a model inferred from actual noncoding regions of the species under consideration should be used instead. However, noncoding DNA is usually under-represented in public databases and it may exhibit a high degree of heterogeneity along the genome. Therefore, we believe that, at least for illustration purposes, the random assumption does not introduce a significant distortion.

## 4.2.1   Measures based on oligonucleotide counts

Unequal usage of codons in the coding regions appears to be a universal feature of the genomes across the phylogenetic spectrum. This bias obeys mainly (1) the uneven usage of the amino acids in the existing proteins and (2) the uneven usage of synonymous codons (Grantham *et al.*, 1980). The bias in the usage of the synonymous codons could correlate with the abundance of the corresponding tRNAs in some organisms (Ikemura, 1985), although this is unclear.

### 4.2.1.1   Codon usage

By comparing the frequency of codons in a region of a species genome read in a given frame with the typical frequency of codons in the species genes, it is possible to estimate a likelihood of the region coding for a protein in such a

**Table 4.1**  The human codon usage and codon preference table as published in http://bioinformatics.weizmann.ac.il/databases/codon. For each codon, the table displays the frequency of usage of each codon (per thousand) in human coding regions (first column) and the relative frequency of each codon among synonymous codons (second column).

**The human codon usage table**

| | | | | | | | | | | | | | | | |
|---|---|---|---|---|---|---|---|---|---|---|---|---|---|---|---|
| Gly | GGG | 17.08 | 0.23 | Arg | AGG | 12.09 | 0.22 | Trp | TGG | 14.74 | 1.00 | Arg | CGG | 10.40 | 0.19 |
| Gly | GGA | 19.31 | 0.26 | Arg | AGA | 11.73 | 0.21 | End | TGA | 2.64 | 0.61 | Arg | CGA | 5.63 | 0.10 |
| Gly | GGT | 13.66 | 0.18 | Ser | AGT | 10.18 | 0.14 | Cys | TGT | 9.99 | 0.42 | Arg | CGT | 5.16 | 0.09 |
| Gly | GGC | 24.94 | 0.33 | Ser | AGC | 18.54 | 0.25 | Cys | TGC | 13.86 | 0.58 | Arg | CGC | 10.82 | 0.19 |
| Glu | GAG | 38.82 | 0.59 | Lys | AAG | 33.79 | 0.60 | End | TAG | 0.73 | 0.17 | Gln | CAG | 32.95 | 0.73 |
| Glu | GAA | 27.51 | 0.41 | Lys | AAA | 22.32 | 0.40 | End | TAA | 0.95 | 0.22 | Gln | CAA | 11.94 | 0.27 |
| Asp | GAT | 21.45 | 0.44 | Asn | AAT | 16.43 | 0.44 | Tyr | TAT | 11.80 | 0.42 | His | CAT | 9.56 | 0.41 |
| Asp | GAC | 27.06 | 0.56 | Asn | AAC | 21.30 | 0.56 | Tyr | TAC | 16.48 | 0.58 | His | CAC | 14.00 | 0.59 |
| Val | GTG | 28.60 | 0.48 | Met | ATG | 21.86 | 1.00 | Leu | TTG | 11.43 | 0.12 | Leu | CTG | 39.93 | 0.43 |
| Val | GTA | 6.09 | 0.10 | Ile | ATA | 6.05 | 0.14 | Leu | TTA | 5.55 | 0.06 | Leu | CTA | 6.42 | 0.07 |
| Val | GTT | 10.30 | 0.17 | Ile | ATT | 15.03 | 0.35 | Phe | TTT | 15.36 | 0.43 | Leu | CTT | 11.24 | 0.12 |
| Val | GTC | 15.01 | 0.25 | Ile | ATC | 22.47 | 0.52 | Phe | TTC | 20.72 | 0.57 | Leu | CTC | 19.14 | 0.20 |
| Ala | GCG | 7.27 | 0.10 | Thr | ACG | 6.80 | 0.12 | Ser | TCG | 4.38 | 0.06 | Pro | CCG | 7.02 | 0.11 |
| Ala | GCA | 15.50 | 0.22 | Thr | ACA | 15.04 | 0.27 | Ser | TCA | 10.96 | 0.15 | Pro | CCA | 17.11 | 0.27 |
| Ala | GCT | 20.23 | 0.28 | Thr | ACT | 13.24 | 0.23 | Ser | TCT | 13.51 | 0.18 | Pro | CCT | 18.03 | 0.29 |
| Ala | GCC | 28.43 | 0.40 | Thr | ACC | 21.52 | 0.38 | Ser | TCC | 17.37 | 0.23 | Pro | CCC | 20.51 | 0.33 |

frame. Regions in which codons are used with frequencies similar to the typical species codon frequencies are likely to code for genes. This idea was first introduced by Staden & McLachlan (1982). In the practice, the likelihood can be computed in a number of different ways. Here, we compute it as a log-likelihood ratio. Let $F(c)$ be the frequency (probability) of codon $c$ in the genes of the species under consideration (in other words, $F$ is the codon usage table, see Table 4.1). Then, given a sequence of codons $C = C_1 C_2 \ldots C_m$, and assuming independence between adjacent codons

$$P(C) = F(C_1)\, F(C_2) \ldots F(C_m)$$

is the probability of finding the sequence of codons $C$ knowing that $C$ codes for a protein. For example, if $S$ is the sequence $S = \text{AGGACG}$, when read in frame 1, it results in the sequence $C^1_1 = \text{AGG}$, $C^1_2 = \text{ACG}$. Then

$$P^1(S) = P(C^1) = F(\text{AGG})\, F(\text{ACG})$$

Substituting the appropriate values from Table 4.1, we obtain

$$P^1(S) = P(C^1) = 0.022 \times 0.038 = 0.000836$$

Alternatively, let $F_0(c)$ be the frequency of codon $c$ in a noncoding sequence.

$$P_0(S) = P_0(C) = F_0(C_1) F_0(C_2) \ldots F_0(C_m)$$

is the probability of finding the sequence $S$ if $C$ is noncoding. Assuming the random model of coding DNA, $F_0(c) = 1/64 = 0.0156$ for all codons, and $P_0$ for the above sequence of codons $C$ would be

$$P_0(C) = 0.0156 \times 0.0156 = 0.000244$$

The log-likelihood ratio for $S$ coding in frame 1, $LP^1$, is

$$LP^1(S) = \log(0.000836/0.000244) = \log(3.43) = 0.53$$

The log-likelihood ratios for $S$ coding in frames 2 and 3 ($LP^2$ and $LP^3$) are computed in a similar way. Table 4.2 shows the values of the log-likelihood ratios computed on our test exon and intron sequences, using the values of $F$ from Table 4.1. As it can be seen in this case the log-likelihood ratio $LP$ is indeed greater than zero in the coding frame of the exon sequence, while it is smaller than zero in the noncoding frames of the exon sequence and in all frames of the intron sequence.

The distribution of the scores of the codon usage log-likelihood ratios in the larger sets of intron and exon sequences are shown in Fig. 4.1. Because the sequences in these sets are of very different lengths, the scores are divided by the sequence length in order to derive these distributions. It can be seen that, although the distributions are clearly distinct, there is substantial overlap between the codon usage scores in the sets of intron and exon sequences. As will be shown, this is a general situation for all coding statistics.

In practice, the problem is not usually to determine the likelihood that a

**Table 4.2**   Values of different coding statistics in the 223 bp second coding exon of the human β-globin gene, and in a 223 bp sequence from the middle of the second intron of the same gene.

| | Exon sequence | | | Intron sequence | | |
|---|---|---|---|---|---|---|
| | Coding frame | Noncoding frames | | Frame 1 | Frame 2 | Frame 3 |
| Codon usage | 24.06 | −16.13 | −3.16 | −14.36 | −23.74 | −19.67 |
| Hexamer usage | 27.62 | −11.64 | −6.51 | −20.90 | −27.56 | −22.07 |
| | 39.98 | −14.58 | −8.46 | −26.73 | −27.81 | −25.87 |
| Codon preference | 15.97 | −1.32 | 7.24 | −7.96 | −12.70 | −14.93 |
| Amino acid usage | 8.17 | −14.87 | −10.17 | −6.15 | −10.69 | −4.57 |
| Codon prototype | 9.87 | −11.23 | −10.30 | −11.45 | −17.44 | −14.49 |
| Markov model | | | | | | |
| order 1 | 29.92 | −2.69 | −3.31 | −35.44 | −42.40 | −41.73 |
| order 2 | 34.73 | −18.26 | −7.77 | −29.61 | −41.76 | −40.05 |
| order 5 | 72.69 | −21.38 | 13.56 | −37.63 | −30.99 | −36.40 |
| Position Asymmetry | 0.0957 | | | 0.0211 | | |
| Periodic Asymmetry Index | 1.159 | | | 1.009 | | |
| Average Mutual Information | 0.00681 | | | 0.000344 | | |
| Fourier spectrum | 2.278 | | | 0.892 | | |

given sequence is coding or not, but to locate the (usually small) coding regions within large genomic sequences. The typical procedure is to compute the value of a coding statistic in successive (usually overlapping) windows (a sliding window) and record the value of the statistic for each of the windows. This generates a profile along the sequence in which peaks may point to the coding regions and valleys to the noncoding regions. Figure 4.2 plots the result of sliding a window of 120 bp (the distance between consecutive windows being 10 bp), computing *LP* in the three different frames, and plotting the highest value obtained. In this case, the resulting profile reproduces fairly well the exonic structure of the human β-globin gene.

As already stated, codon usage bias is a mixture of both bias in the usage of amino acids and bias in the usage of synonymous codons. Methods can be used to measure these effects separately.

## 4.2.1.2   Amino acid usage

McCaldon & Argos (1988) computed the probabilities of occurrence of different oligopeptides in existing proteins. By translating a sequence of codons to a sequence of amino acids, the probability of the resulting sequence of oligopeptides assuming the region to be coding can be computed. Here, following

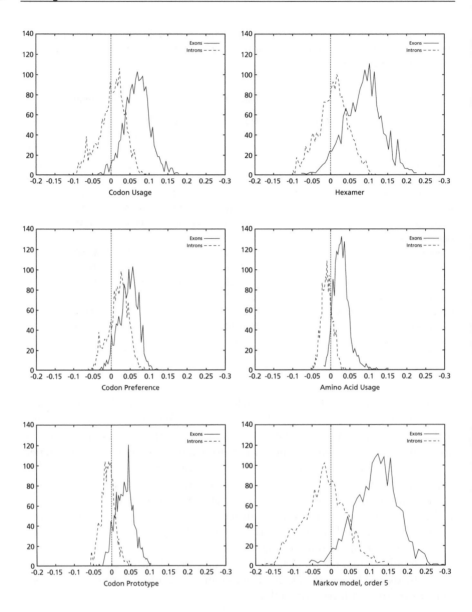

**Figure 4.1**    Distribution of the scores of the model based Coding Statistics in the set of 1761 human exons and 1753 human intron sequences. To plot them, the values of the coding statistics are divided by the length of the sequence.

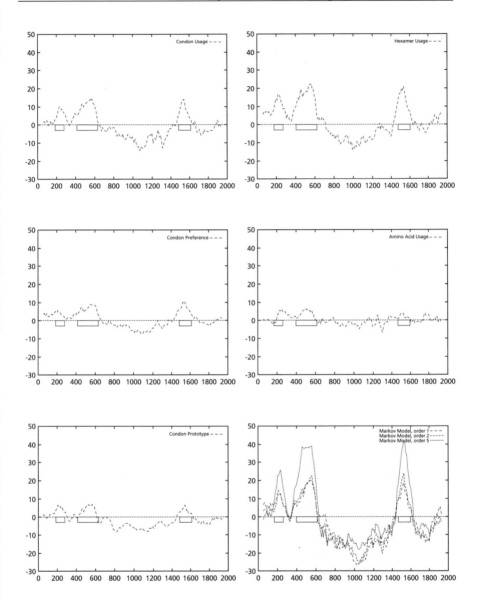

**Figure 4.2** Values of the model based coding statistics along the 2000 bp human β-globin gene sequence, computed on a sliding window of length 120 and step 10. Boxes represent the positions of the coding exons along the sequence.

Fickett & Tung (1992), a measure of amino acid bias is computed based on the observed frequencies of single amino acids in the existing proteins. The measure is identical to the log-likelihood ratio introduced to measure codon usage bias, but the probability of each codon is now the observed probability of the amino acid encoded by the codon. That is, $F_A(C)$ is the observed probability of the amino acid encoded by the codon $c$ in the existing proteins. This value can be directly derived from a codon usage table by summing up the probabilities of synonymous codons; that is, given a codon $c$

$$F_A(c) = \sum_{c' \equiv c} F(C')$$

where, $c' \equiv c$ means $c'$ synonymous to $c$. Then,

$$P^i_A(S) = P_A(C^i) = F_A(C^i_1)F_A(C^i_2) \ldots F_A(C^i_m)$$

is the probability of finding the sequence of amino acids resulting of translating the sequence $S$ in frame $i$ given that $S$ is coding in frame $i$. As a model of noncoding DNA, we assume the probability of each amino acid to be proportional to the number of synonymous codons coding for the amino acid, that is $F_{A0}(C) = n_c/64$, where $n_c$ is the number of codons synonymous to $c$, and we can compute $P_{A0}(S)$. Then, from $P^i_A(S)$ and $P_{A0}(S)$, we compute the amino acid usage log-likelihood ratio as usual. Results obtained using this measure in our test sequences are shown in Table 4, and Figures 1 and 2.

### 4.2.1.3 Codon preference

Gribskov et al. (1984) introduce a coding statistic to measure uneven usage of synonymous codons solely. Indeed, from a codon usage table, we can compute the relative probability of each synonymous codon to code for a given amino acid. For example, from Table 4.1, we can see that codons GAG and GAA (the two codons coding for glutamic acid) are used in coding regions with probabilities 0.03882 and 0.02751, respectively, which results in a relative probability of 0.59 and 0.41, respectively. Let $F_R(c)$ be the relative probability in coding regions of codon $c$ among codons synonymous to $c$,

$$F_R(C) = \frac{F(C)}{\sum\limits_{c' \equiv c} F(C')}$$

Then

$$P^i_R(S) = P_R(C^i) = F_R(C^i_1)F_R(C^i_2) \ldots F_R(C^i_m)$$

is the probability of the sequence $S$ given the particular sequence of amino acids coded by $S$ in frame $i$ (that is, in $P_R$ the effects of unequal usage of amino acids have been eliminated.) We will assume that in noncoding DNA, there is no preference between synonymous codons to code for a given amino acid; therefore, the probability of codon $c$ in non-coding DNA is $F_{R0} = 1/n_c$. From $P^i_R(S)$ and $P_{R0}(S)$ we compute the codon preference log-likelihood ratio as usual. Results obtained using this measure in our test sequences are shown in Table 4.2, and Figs 4.1 and 4.2.

As can be seen from Table 4.2 and Figure 4.2, although amino acid usage and codon preference carry a lot of information about coding function, neither of these measures appears to be as discriminant as codon usage. It is easy to see that, as we have introduced them, codon usage is the combination of amino acid usage and codon preference. Indeed, from the definitions above, it follows directly that for a given codon:

$$F(C) = F_A(C) \, F_R(C)$$
$$F_0(C) = F_{A0}(C) \, F_{R0}(C)$$

which results in

$$P^i(S) = P^i_A(S)P^i_R(C)$$
$$P_0(S) = P_{A0}(C)P_{R0}(C)$$

for a sequence $S$ in frame $i$, which in turn leads to

$$LP^i(S) = LP^i_A(S) + LP^i_R(S)$$

which states that codon usage bias is the sum of amino acid usage bias and codon preference.

## 4.2.1.4   Hexamer usage

Bias in the distribution of oligonucleotides other than codons (trinucleotides) can also be used to discriminate between coding and noncoding regions. Bias in the usage of hexamers may be the most discriminant one (probably because of dependence between adjacent amino acids in the proteins). Claverie *et al.* (1990) were the first to use hexamer frequencies to locate coding regions. Bias in hexamer usage can be computed exactly as bias in codon usage. A hexamer usage table, $F(H_i)$ ($i = 1, \ldots, 4096$) from the species under consideration is computed a priori. The probability of a sequence of hexanucleotides, $H = H_1,$ $H_2, \ldots, H_m$, in the coding frame of a coding sequence is $P(H) = F(H_1) \, F(H_2) \cdots F(H_m)$. If $P_0$ is the background probability distribution, the log-likelihood ratio $LP$ can be computed as before. A test sequence can now be decomposed in six different sequences of hexamers, instead of three, and, thus, six log-likelihood ratios can be computed ($LP^i$, $i = 1 \ldots 6$). Table 4.2 shows the values of these ratios in our test intron and exon sequences, Fig. 4.1 shows the distributions of the standardized scores (by the sequence length) in the larger sets of intron and exon sequences, and Fig. 4.2 shows the results of sliding a window and plotting the maximum of the six values at each position.

## 4.2.2 Measures based on base compositional bias between codon positions

From the codon usage table (Table 4.1), we can derive the probability of each base at each codon position in coding regions (Table 4.3). It can be seen that there are clear differences in the frequency with which the different bases appear at the different codon positions; for example, G is almost twice as frequent as T in the first codon position, while T is more frequent than G in the second codon position. Similarly, C is almost twice as frequent as A in the third codon position, but A is more frequent than C in the second codon position. Shepherd (1981) noted that the most frequent codons were of the form RNY (R = A or G, Y = C or T, N any nucleotide). He suggested a method to test for the existence and frame of a coding region by measuring the number of differences between the sequence and the pattern RNYRNRY . . . RNY. A number of other measures have since been proposed to exploit the asymmetry in the base composition between codon positions in order to locate potential coding regions in genomic DNA.

Before discussing some of these measures, we would like to point out that asymmetry in the base composition between codon positions arises, not only because of uneven usage of amino acids and synonymous codons, but also because of the particular structure of the genetic code. Indeed, uneven usage of amino acids and uneven usage of synonymous codons are not enough to produce asymmetry in base composition, as the example illustrates on page 65.

Because of the structure of the genetic code, synonymous codons almost always share the first two nucleotides (the exception being obviously the amino acids coded by six codons, arginine, serine, and leucine). This implies that the first two positions in the codons will be more abundant in those nucleotides common to the synonymous codons corresponding to the most abundant

**Table 4.3**  Frequency of the four different nucleotides at the three different codon positions in human coding regions. Derived from Table 4.4.

| Nucleotide | Codon position | | |
|---|---|---|---|
| | 1 | 2 | 3 |
| A | 0.27 | 0.31 | 0.18 |
| C | 0.24 | 0.24 | 0.31 |
| G | 0.32 | 0.20 | 0.29 |
| T | 0.17 | 0.26 | 0.22 |

## EXAMPLE. CODON PREFERENCE AND AMINO ACID USAGE BIAS DO NOT NECESSARILY RESULT IN CODON ASYMMETRY IN BASE COMPOSITION

Let's assume a three letter 'DNA' and 'amino acid' codes:

$$DNA = \{A, B, C\} \text{ and } AA = \{P, Q, R\}$$

Let's assume 'codons' to be di-nucleotides, and let's assume strong bias in 'amino acid' usage, and in 'codon' preference, as expressed in the following 'genetic code' table:

| P | 0.7 |
|---|-----|
| Q | 0.2 |
| R | 0.1 |

amino acid usage bias

| P | AA | 0.6 | Q | BB | 0.6 | R | AB | 0.5 |
|---|----|-----|---|----|-----|---|----|-----|
| P | BC | 0.2 | Q | AC | 0.2 | Q | BA | 0.5 |
| P | CB | 0.2 | Q | CA | 0.2 | stop | CC | |

Codon preference bias

This results in a very biased 'codon' usage table, but not in 'codon' asymmetry in base composition

| AA | 0.42 | AB | 0.05 | AC | 0.04 |
|----|------|----|------|----|------|
| BA | 0.05 | BB | 0.12 | BC | 0.14 |
| CA | 0.04 | CB | 0.14 | CC | |

'codon' usage

| | Codon position | |
|---|---|---|
| | 1 | 2 |
| A | 0.51 | 0.51 |
| B | 0.31 | 0.31 |
| C | 0.18 | 0.18 |

base composition at 'codon' positions

amino acids. Conversely, differences between synonymous codons are confined mostly to the third codon position. At this position, C and G are usually preferred, as genes (at least, in higher eukaryotes) tend to occur in G+C rich regions (Marin *et al.* 1989).

## 4.2.2.1 Codon prototype

The distribution of base frequencies at codon positions (Table 4.3) can be assumed to describe statistically a prototypical codon. Then, given a sequence problem $S$, we can measure how similar to the prototypical distribution is the observed distribution of base frequencies at the three codon positions in $S$. The closer the distributions the more likely it is for $S$ to be coding. As usual, there are a number of ways in which such a proximity can be measured (Mural *et al.*, 1991; Fickett & Tung, 1992). Here, we will compute the usual log-likelihood ratio.

Let $f(b, r)$ be the probability of nucleotide $b$ at codon position $r$, as estimated from known coding regions (Table 2). Then, if $C$ is a codon

$$F(C) = f(C[1], 1)f(C[2], 2)f(C[3], 3)$$

is the probability of codon $c$ in coding regions, assuming independence between adjacent nucleotides. On the other hand, we will assume $F_0(c) = 1/64$ the probability of all triplets $c$ in noncoding DNA. From $F$ and $F_0$, $P^i$ and $P_0$ can be computed exactly as before when deriving the codon usage log-likelihood ratio. For example, if the sequence $S$ is AGGACG, the probability of $S$, if $S$ is coding in frame 1, can be computed as

$$P^1(S) = F(AGG)F(ACG) = f(A, 1)\, f(G, 2)\, f(G, 3)\, f(A, 1)\, f(C, 2)\, f(G, 3)$$

From Table 4.3, we obtain

$$P^1(S) = \underbrace{0.27 \times 0.20 \times 0.29}_{F(AGG) = 0.01566} \times \underbrace{0.27 \times 0.24 \times 0.29}_{F(ACG) = 0.01879} = 0.0002943$$

$P^2(S)$ and $P^3(S)$ are computed in a similar way. From $P^1$, and $P_0$, the codon preference log-likelihood ratio is derived as usual. Results obtained using this measure are shown in Table 4.2, and Figures 4.1 and 4.2.

## 4.2.3    Measures based on dependence between nucleotide positions

Both codon prototype and codon usage are based on a model of coding DNA described by the probabilities of the codons. The models, however, are very different. In codon usage, the model is described by the explicit probability of each codon. In codon prototype, the model is described simply by the probability of occurrence of each base at each position in a codon. Codon prototype and codon usage would be equivalent if codon positions were independent. This is not clearly the case: the frequencies of codons derived from Table 4.3 (assuming independence between codon positions) are substantially different from the observed codon frequencies (Table 4.1). Measures based on the frequency of usage of oligonucleotides, such as codon usage, implicitly capture such dependences between nucleotide positions within codons in coding regions. Dependencies between nucleotide positions in coding regions, however, can be explicitly described by means of Markov models.

## 4.2.3.1   Markov models

Borodovsky & McIninch (1993) first introduced the usage of Markov models to locate potential coding regions in DNA sequences. For illustration purposes, it may be helpful to introduce Markov models from codon prototype. As we have seen, in codon prototype, the probability of a nucleotide appearing in a given codon position is constant, independent of the nucleotides in nearby positions. For example, if $S$ is the sequence above ($S$ = AGGACG), the probability of G at codon position 3 ($S_3$ and $S_6$) is constant, 0.29, whether the nucleotide preceding G is G (as in $S_2$) or C (as in $S_5$). In the Markov models, however, the probability of a nucleotide at a particular codon position depends (is conditioned) on (by) the nucleotide(s) preceding it.

In the simplest of the Markov models, the order 1 Markov models, the probability of a nucleotide depends only on the preceding nucleotide. In this case, the coding DNA model is based on the probabilities of the four nucleotides at each codon position, depending on the nucleotide occurring at the preceding codon position (called the *transition probabilities*). Thus, instead of one single matrix, as in codon prototype, three $4 \times 4$ matrices (the *transition matrices*) are required, $F^1$, $F^2$, and $F^3$ each corresponding to a different codon position. Coefficient $i, j$ from matrix $F^r$, $F^r(i, j)$, corresponds to the probability of nucleotide $i$ in codon position $r + 1$ (position 1, if $r$ = 3), given that nucleotide $j$ is at codon position $r$. We have estimated these matrices from the sample of 1761 human exons. The conditional probability of nucleotides $i$ in codon position $r + 1$, given nucleotide $j$ in codon position $r$, is estimated by the number of times that dinucleotide $j, i$ appears at codon position $r$ over the total number of times that nucleotide $j$ appears at codon position $r$. These matrices are shown in Table 4.4. Indeed, we can see from them that the probability of G at codon position 3, given that C is at codon position 2, is 0.27, but the probability of G at codon position 3 given that G is at codon position 2 is 0.37.

The probability of $S$ above, given that $S$ is coding in frame 1, can now be computed as

$$P^1(S) = f(A, 1)F^1(G, A)F^2(G, G)F^3(A, G)F^1(C, A)F^2(G, C)$$

The probability of the first nucleotide in the sequence is not described by the transition matrices because it is not preceded by any nucleotide. For the probability of the first nucleotide in the sequence (the *initial probability*) we can simply assume the probability of such nucleotide depending on its codon position (that is, the value in the codon prototype table; Table 4.3). Then, substituting the appropriate values from Tables 4.3 and 4.4 in the above equation, we obtain the probability of $S$ coding in frame 1:

$$P^1(S) = 0.27 \times 0.19 \times 0.27 \times 0.24 \times 0.21 \times 0.41 = 0.0002862$$

Similarly, the probability of the sequence of nucleotides $S$, given that $S$ codes in frame 2 is

**Table 4.4** Probabilities of the four nucleotides at the different codon positions conditioned to the nucleotide in the preceding codon position. Estimated from our set of human exon and intron sequences.

|                  | A    | C    | G    | T    |
|------------------|------|------|------|------|
|                  | Codon position 3 | | | |
| Codon position 1 |      |      |      |      |
| A                | 0.22 | 0.33 | 0.24 | 0.13 |
| C                | 0.21 | 0.29 | 0.27 | 0.21 |
| G                | 0.44 | 0.15 | 0.37 | 0.53 |
| T                | 0.13 | 0.22 | 0.12 | 0.13 |
|                  | Codon position 1 | | | |
| Codon position 2 |      |      |      |      |
| A                | 0.36 | 0.27 | 0.35 | 0.18 |
| C                | 0.21 | 0.23 | 0.24 | 0.27 |
| G                | 0.19 | 0.14 | 0.23 | 0.23 |
| T                | 0.24 | 0.35 | 0.19 | 0.31 |
|                  | Codon position 2 | | | |
| Codon position 3 |      |      |      |      |
| A                | 0.16 | 0.19 | 0.15 | 0.07 |
| C                | 0.28 | 0.44 | 0.41 | 0.33 |
| G                | 0.40 | 0.12 | 0.27 | 0.45 |
| T                | 0.16 | 0.25 | 0.17 | 0.16 |

$$P^2(S) \quad = \quad f(A, 2) \quad F^2(G, A) \quad F^3(G, G) \quad F^1(A, G) \quad F^2(C, A) \quad F^3(G, C)$$

$$\| \qquad\qquad \| \qquad\quad \| \qquad\quad \| \qquad\quad \| \qquad\quad \|$$

$$0.0006744 \quad = \quad 0.31 \quad\; 0.40 \quad\;\; 0.37 \quad\;\; 0.35 \quad\;\; 0.28 \quad\;\; 0.15$$

$P^3(S)$ can be computed in exactly the same way. If $F_0$ is the matrix of conditional probabilities of the different nucleotides given the preceding ones, we can compute the usual log-likelihood ratio. Assuming a random model of noncoding DNA, the probability of a nucleotide $i$ does not depend on the preceding nucleotide $j$, then $F_0(i, j) = 0.25$ for each pair of nucleotides $i$, $j$, Results obtained in the test sequences with the log-likelihood ratios corresponding to a Markov model of order 1 are shown in Table 4.2, and Figures 4.1 and 4.2.

In Markov models of order 2, the probability of a given nucleotide at a given codon position depends on the dinucleotide preceding it. The transition matrices have now four rows and 16 columns (one for each possible dinucleotide). $F^2$ (A, GC), for example, would be the probability of A following the dinucleotide GC at codon position 2 (that is, A would be at codon position 1). In general, the order of the Markov model indicates the number of preceding nucleotides on which the probability of a given nucleotide depends. In a Markov model of order $k$, thus, the coefficients $F_k^r(i, j)$ correspond to the probability of oligonucleotide $j$ of length $k + 1$ at codon position $r$, given oligonucleotide $i$ of length $k$ at codon position $r$. These probabilities are estimated by the frequency of oligonucleotide $j$ of length $k + 1$ at codon position $r$ over the frequency of oligonucleotide $j$ of length $k$ at position $r$. Borodovsky & McIninch (1993) investigated Markov models of order up to $k = 5$. Results obtained in the test sequences with Markov models of order 2 and 5 are shown in Table 4.2, and Figs 4.1 and 4.2.

Markov models of higher order may capture more of the intrinsic features of coding DNA, but they also depend on more parameters. Since Markov models of order 2 are based on counts of trinucleotides they are similar to codon usage. However, while codon usage reflects only dependences between contiguous nucleotides within codons, a Markov model of order 2 also reflects nucleotide dependences between nucleotides in contiguous codons. Consequently, it also depends on more parameters: while codon usage depends on 64 probabilities, a Markov model of order 2 depends on four $4 \times 16$ matrices (corresponding to the three transition matrices and the initial probabilities matrix), that is, it depends on 256 estimated probabilities. A similar reasoning can be applied to the relationships between hexamer usage and Markov models of order 5.

## 4.3   MEASURES INDEPENDENT OF A MODEL OF CODING DNA

All the methods reviewed so far rely on a probabilistic model for the contents of coding DNA, under which the coding likelihood of DNA sequences is computed. To estimate the probabilities describing the coding model (of codons, amino acids, synonymous codons, hexamers, nucleotides at codon positions, etc.) a non-biased sample of coding DNA is ideally required. However, for most species such a sample does not exist. Indeed, for most eukaryotic organisms (other than *Saccharomyces cerevisiae* and *Caenorhabditis elegans*) only a small fraction of the genes are known. The set of known genes tend to be biased towards those that are highly expressed and could exhibit characteristic sequence features, such as strong codon preference bias. The situation is even worst in the case of the prokaryotic genomes. Recent technological progress has made shotgun sequencing of whole prokaryotic genomes (up to a few megabases) feasible. It is usually the case, therefore, that no sequences of genes are known before the whole genome of a prokaryotic organism is sequenced. Coding measures not depending on an a priori model of coding DNA would, therefore, be very useful. A number of such measures has been proposed. In general, the underlying assumption is that coding DNA is less random or homogeneous than noncoding DNA. Deviation from randomness or inhomogeneity can be measured independently of a reference model and the resulting score correlated with coding function. Since there is no reference model, these scores do not have a direct probabilistic meaning, although their distribution can be studied empirically in known sets of coding and noncoding sequences.

Deviation from randomness or inhomogeneity may in practice mean a number of different things. Fichant & Gautier (1987) used correspondence analysis to measure the degree of homogeneity in codon usage between the three frames of the sequence. The assumption is that if the sequence is coding, codon usage will be markedly different in the coding frame than in the two other frames. Therefore there will exist inhomogeneity in codon usage between frames, while if the sequence is not coding, codon usage will be essentially the same in the three frames and codon usage will be homogeneous between frames. While the Fichant and Gautier (1987) measure is based on the usage of codons, Fickett (1982) and Staden (1984) proposed measures independent of a reference model, based on the asymmetry in the base composition between codon positions. We discuss one such measure next.

### 4.3.1   Measures based on base compositional bias between codon positions

#### *4.3.1.1   Position asymmetry*

The goal here is to measure how asymmetric is the distribution of nucleotides at the three triplet positions in the sequence problem. Fickett (1982) and Staden (1984) calculated the asymmetry independently for each nucleotide (although using different formulae), and then combined the values into a single score. Staden (1984) simply summed the four values. Fickett (1982), in the widely used TEST-CODE program, considered in addition the frequencies of the four nucleotides. Each of the asymmetries and frequencies is used to make an estimate of coding likelihood that makes this measure dependent on a sample set of known coding and noncoding sequences. The separate estimates are then combined using a linear weighted sum. Here, we also compute the asymmetry independently for each nucleotide, simply as the variance of the frequency of the nucleotide at the three codon positions, as suggested in Fickett & Tung (1992), and sum the four values obtained into a single score. Let $f_s(b,r)$ be the (relative) frequency of nucleotide $b$ at codon position $r$ in the sequence problem $S$, as calculated from one of the three decompositions of $S$ in codons (any of them). Let

$$f_s(b) = \sum_{r=1}^{3} (f_s(b, r))/3$$

be the average frequency of nucleotide $b$ at the three codon positions and define the asymmetry in the distribution of nucleotide $b$, as the variance of this frequency

$$\text{asym}(b) = \sum_{i=1}^{3} (f_S(b, i) - f_s(b))^2.$$

Note that the value of asym($b$) is independent of the frame in $S$ in which the codons are defined. Therefore, only one value of asymmetry needs to be computed along the sequence problem (and not one for each frame, as we have been doing so far). We then compute the position asymmetry of the sequence, $PA(S)$ as

$$PA(S) = \text{asym}(A) + \text{asym}(C) + \text{asym}(G) + \text{asym}(T)$$

Table 4.2 and Figs 4.3 and 4.4 show the results obtained when PA is computed in our test sequences.

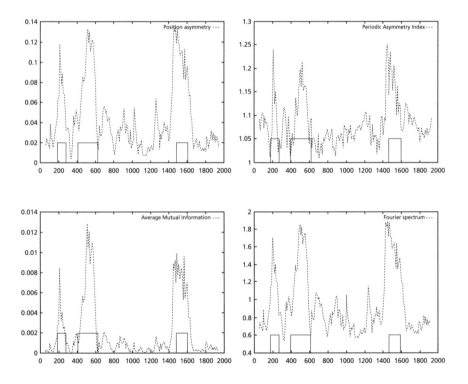

**Figure 4.3**   Values of the model-independent coding statistics along the 2000 bp human β-globin gene sequence, computed on a sliding window of length 120 and step 10.

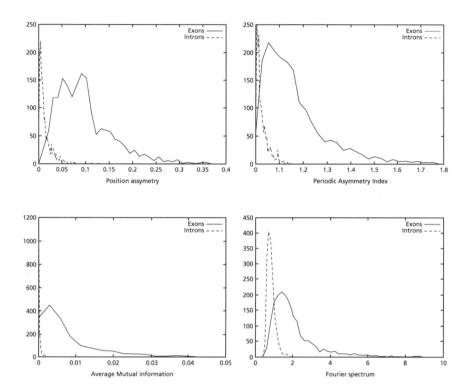

**Figure 4.4**    Distribution of the scores of the model independent coding statistics in the set of 1761 human exons and 1753 human intron sequences. To plot them, the values of the coding statistics were divided by the length of the sequence.

## 4.3.2 Measures based on periodic correlations between nucleotide positions

Given a DNA sequence, we can compute how many times nucleotide $i$ is followed by nucleotide $j$ at a distance of $k$ nucleotides, $N_{ij}(k)$. For example, if the sequence

$$S = AGGACGGGATCA$$

then $N_{GA}(1) = 2$, $N_{AT}(0) = 1$, $N_{GG}(0) = 3$, $N_{AA}(7) = 2$, and so on. Figure 4.5 shows the absolute frequency of the pair A . . . A with $k$ nucleotides between the two As occurring in the 200 first base pairs of the sequences in our test sets of human exons and introns. A clear periodic pattern arises from the set of

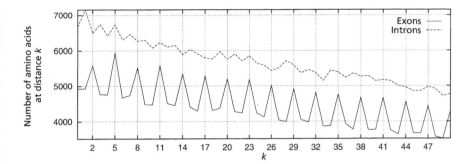

**Figure 4.5** Periodic structure in DNA sequences. The absolute frequency of the pair A . . . A with $k$ (from 0 to 50) nucleotides between the two As in the 200 first base pairs of the sequences in the set of 1761 human exons and 1753 human introns. A clear period-3 pattern appears in coding regions, which is absent in noncoding regions. Owing to the finite size of the sequences (200 bp) the periodic pattern vanishes at longer distances $k$. A similar periodic pattern appears in coding regions for the other 15 possible pairs of nucleotides.

exon sequences. The nucleotide A is more likely to be found at distance $k = 2$, 5, 8, . . . from another A than at other distances. Note that nucleotide pairs at a distance of $k = 2$, 5, 8, . . . nucleotides, are at the same codon position, whereas nucleotide pairs at other distances, are not. Such a periodic pattern reflects correlations between nucleotide positions along coding sequences (that is, the probability of finding a nucleotide at a given position in a coding sequence is not independent of the nucleotide occurring at some other – even distant – position). The correlations arise, in turn, because of the asymmetry in base composition at the three codon positions in coding sequences (Gutiérrez *et al.*, 1994). The periodic pattern, which is characteristic of the 16 pairs of nucleotides, and not only of the pair A . . . A, is absent in the intronic sequences.

A number of coding statistics have been devised based on measuring the periodic structure (or the correlation structure) of DNA sequences. We discuss three such measures next. Konopka (1994), in the 'position asymmetry index', compared the probability that pairs of the same nucleotide appear at a distance $k = 2$, 5, 8 . . . (that is, at the same codon position) in the query sequence with the probability that these pairs appear at other distances (different codon positions). Herzel & Grosse (1995), in the 'Average Mutual Information', compared the correlations of all pairs of nucleotides at the same codon positions with the correlations at different codon positions. Finally, Tiwari *et al.* (1997) used the relative peak at the frequency 1/3 in the Fourier spectrum of the sequence.

## 4.3.2.1 The periodic asymmetry index

Given a sequence $S$, Konopka (1994) considered three distinct probabilities: the probability $P_{[in]}$ of finding pairs of the same nucleotide at distances $k = 2, 5, 8, \ldots$, the probability $P_{out}$ of finding pairs of the same nucleotide at distances $k = 0, 3, 6, \ldots$, and the probability $P^2_{out}$ of finding pairs of the same nucleotide at distances $k = 1, 4, 7, \ldots$ Because of the three-base periodic pattern, in coding regions $P_{[in]}$ will be larger than the other two probabilities, while in noncoding regions the three probabilities will be similar. The tendency to cluster homogeneous dinucleotides in a three-base periodic pattern can be measured by the Periodic Asymmetry Index

$$\text{PAI}(S) = \frac{\max(P_{[in]}, P^1_{out}, P^2_{out})}{\min(P_{[in]}, P^1_{out}, P^2_{out})}$$

which can be taken as an indicative of the coding potential of the sequence $S$. Konopka (1994) computed the PAI in a slightly different way, he computes the tendency to cluster dinucleotides in a two-base periodic pattern (suggested to be characteristic of intronic sequences) and the PAI as the ratio of the two tendencies (two-base over three-base periodicity).

Table 4.2, and Figures 4.3 and 4.4 show the results obtained when PAI is computed in our test sequences.

## 4.3.2.2    Average mutual information

Given a sequence $S$, let $P_{ij}(k)$ be the probability in the sequence of the pair of nucleotides $i$ and $j$ at a distance of $k$ nucleotides. These probabilities can be estimated by the absolute frequencies $N_{ij}(k)$ above – see Li (1997) for considerations regarding the estimation of these probabilities. The correlation between nucleotide $i$ and nucleotide $j$ at a distance of $k$ nucleotides can be calculated (Li, 1997) as:

$$\rho_{ij}(k) = P_{ij}(k) - p_i p_j$$

where $p_i$ and $p_j$ are the probabilities of nucleotides $i$ and $j$ in $S$. Thus, for each distance $k$, 16 different individual correlations can be calculated. A measure that summarizes all individual correlations at a given distance $k$ is the Mutual Information Function (Shannon, 1948):

$$I(k) = \sum_{i,j \in \{A,C,G,T\}} P_{ij}(k) \log \left( \frac{P_{ij}(k)}{p_i p_j} \right)$$

$I(k)$ quantifies the amount of information that can be obtained from one nucleotide about another nucleotide at a distance $k$. Figure 4.6 shows the

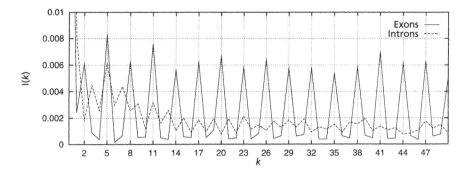

**Figure 4.6**   The Mutual Information Function computed for distances from $k = 0$ to $k = 50$ in the 200 first bp from the sequences in the set of 1761 human exons and in the set of 1753 human introns.

Mutual Information Function computed on the first 200 bp of the sequences in the set of human exons and introns. The three-base periodic pattern in coding sequences becomes obvious. $I(k)$ has larger values for $k = 2, 5, 8, \ldots$ The pattern is absent in noncoding sequences.

In coding DNA $I(k)$ oscillates between two values, while in noncoding DNA $I(k)$ is rather flat. Herzel and Grosse (1995) use this fact to construct a coding statistic. They called the two values between which $I(k)$ oscillates in coding DNA, the in-frame mutual information $I_{in}$ at distances $k = 2, 5, 8, \ldots$, and the out-of-frame mutual information $I_{out}$ at $k = 4, 5, 7, 8, \ldots$ They showed that $I_{in}$ and $I_{out}$ can be computed directly from the probabilities $f_s(b, r)$ of the nucleotide $b$ to appear at codon position $r$, as estimated from $S$. In order to reduce the pair of numbers $I_{in}$ and $I_{out}$ to a single quantity, they computed the Average Mutual Information as (Grosse *et al.*, 1998)

$$AMI = \frac{I_{in} + 2I_{out}}{3}$$

Table 4.2 and Figs 4.3 and 4.4 show the results obtained when AMI is computed in our test sequences.

Correlation structures in DNA sequences, such as those measured in AMI may reveal biologically relevant large-scale heterogeneity in genomic sequences other than coding. For a review of correlation structures in DNA sequences, and their biological implication see Li (1997).

## 4.3.2.3    Fourier spectrum

Periodic correlations in DNA sequences can also be examined by means of Fourier analysis. The partial spectrum of a DNA sequence $S$ of length $l$ corresponding to nucleotide $b$ is defined as (Silverman & Linsker, 1986; Li *et al.*, 1994):

$$S_b(f) = \frac{1}{N^2} \left( \sum_{j=1}^{l} U_b(S_j) e^{2\pi i f j} \right)^2$$

where $U_b(S_j) = 1$ if $S_j = b$, and it is 0 otherwise, and $f$ is the discrete frequency, $f = k/l$, with $k = 1; 2, \ldots, l/2$. The total Fourier spectrum of the DNA sequence is the sum of the four partial spectra:

$$S(f) = \sum_{b \in \{A,C,G,T\}} S_b(f)$$

DNA coding regions reveal the characteristic periodicity of 3 as a distinct peak at frequency $f = 1/3$. No such 'peak' is apparent for noncoding sequences (Tsonis *et al.*, 1991; Tiwari *et al.*, 1997). We have computed the Fourier spectrum at $f = 1/3$ ($S(1/3)$) in our test sequences. Results appear in Table 4.2 and Figs 4.3 and 4.4. It can be seen from Fig 4.3 that the Fourier spectrum profile in the human $\beta$ globin gene sequences is identical (save scale) to the position asymmetry profile. This indicates that there is a one-to-one correspondence between Fourier spectrum and position asymmetry and that one measure can be directly derived from the other. This can be shown analytically (Ivo Grosse, personal communication). The relation between the two measures depends on the length of the sequence, as the dissimilar distributions of the position asymmetry and Fourier spectrum scores indicate in the set of 1761 human exons and 1753 human introns, which have variable length.

In order to obtain a cleaner signal, Tiwari *et al.* (1997) built the ratio of the Fourier spectrum at $f = 1/3$ over the average of the total spectrum of the sequence, ($\bar{S}$), which can be computed from the frequencies of the nucleotides along the sequence.

## 4.4    CODING STATISTICS IN GENE IDENTIFICATION PROGRAMS

A number of gene identification programs for prediction of gene structure in large genomic regions are currently available. Table 4.5 shows a list of available programs and Internet sites to access them. See Fickett (1996), Guigó (1997a) and Claverie (1997) for recent reviews, Burset and Guigó (1996) for a

**Table 4.5**    List of gene identification programs, and Internet access (e-mail server address is provided when different from the WWW address).

| Program | Authors | WWW address |
| --- | --- | --- |
| GENEMODELER | Fields & Soderlund (1990) | |
| GENEID | Guigó *et al.* (1992) | www1.imim.es/geneid.html |
| | | geneid@darwin.bu.edu |
| SORFIND | Hutchinson & Hayden (1992) | |
| GENEPARSER | Snyder & Stormo (1993) | beagle.colorado.edu/ eesnyder/GeneParser.html |
| GENEMARK | Borodovski & McIninch, 1993 | intron.biology.gatech.edu/~genmark |
| | | genmark@ford.gatech.edu |
| GENVIEW | Milanesi *et al.* (1993) | www.itba.mi.cnr.it/webgene |
| GREAT | Gelfand and Roytberg (1993) | |
| GRAIL II/GAP | Xu *et al.* (1994) | avalon.epm.ornl.gov/gallery.html |
| | | grail@ornl.gov |
| FGENEH | Solovyev *et al.* (1994) | dot.imgen.bcm.tmc.edu:9331/gene-finder/gf.html |
| | | service@bchs.uh.edu |
| GENELANG | Dong & Searls (1994) | cbil.humgen.upenn.edu/~sdong/genlang_home.html |
| | | genlang@cbil.humgen.upenn.edu |
| XPOUND | Thomas & Skolnick (1994) | |
| GENIE | Kulp *et al.* (1996) | www-hgc.lbl.gov/inf/genie.html |
| PROCRUSTES | Gelfand *et al.* (1996) | www-hto.usc.edu/software/procrustes/ |
| MZEF | Zhang (1997) | www.cshl.org/genefinder |
| GENSCAN | Burge & Karlin (1997) | gnomic.stanford.edu/GENSCANW.html |
| MORGAN | Salzberg *et al.* (1997) | www.cs.jhu.edu/labs/compbio/morgan.html |
| VEIL | Henderson *et al.* (1997) | www.cs.jhu.edu/labs/compbio/veil.html |

comparative benchmark, and the WWW document maintained by Wential Li at linkage.rockefeller. edu/wli/gene/ for an up-to-date list of references. At the core of all such programs, there exists one or more coding statistics related to the measures reviewed here. Indeed, more powerful programs are entirely built on hidden Markov models (GENSCAN, Burge & Karlin, 1997; GENIE, Kulp *et al.*, 1996; VEIL, Henderson *et al.*, 1997), which are somehow related to the Markov models discussed here.

A general strategy among gene identification programs is to integrate the output of a number of coding statistics. Thus, to name just a few examples, the popular GRAIL program (Uberbacher & Mural, 1991) uses a neural network to integrate a number of coding statistics, mostly related to hexamer usage and position asymmetry in base composition. Solovyev *et al.* (1994) in the FGENEH program use linear discriminant analysis, while Dong & Searls (1994) in GENLANG use linguistic methods. Although increased accuracy in the gene predictions is obtained in this way, because coding statistics are all essentially measuring codon usage bias in one way or another, their output is

**Table 4.6**  Correlation between the different coding statistics in our set of exonic (upper triangle of the table) and intronic (lower triangle) sequences. The scores of the model-dependent coding statistics and of the Fourier Spectrum have been divided by the length of the sequence to compute the correlations.

| | CU | HU | CPre | AAU | CPro | MM-1 | MM-2 | MM-5 | PA | PAI | AMI | FOU |
|---|---|---|---|---|---|---|---|---|---|---|---|---|
| Codon usage | | 0.908 | 0.769 | 0.492 | 0.803 | 0.876 | 0.909 | 0.772 | 0.590 | 0.558 | 0.529 | 0.562 |
| Hexamer usage | 0.925 | | 0.822 | 0.287 | 0.802 | 0.912 | 0.928 | 0.869 | 0.585 | 0.544 | 0.510 | 0.559 |
| Codon preference | 0.927 | 0.932 | | −0.069 | 0.637 | 0.833 | 0.838 | 0.723 | 0.456 | 0.415 | 0.421 | 0.438 |
| Amino acid usage | 0.738 | 0.626 | 0.537 | | 0.351 | 0.238 | 0.262 | 0.199 | 0.392 | 0.391 | 0.355 | 0.383 |
| Codon prototype | 0.822 | 0.820 | 0.782 | 0.623 | | 0.810 | 0.799 | 0.673 | 0.708 | 0.659 | 0.611 | 0.662 |
| Markov model $k = 1$ | 0.943 | 0.941 | 0.952 | 0.604 | 0.875 | | 0.969 | 0.801 | 0.543 | 0.502 | 0.465 | 0.512 |
| Markov model $k = 2$ | 0.974 | 0.944 | 0.952 | 0.665 | 0.853 | 0.976 | | 0.831 | 0.535 | 0.494 | 0.459 | 0.507 |
| Markov model $k = 5$ | 0.919 | 0.932 | 0.913 | 0.621 | 0.816 | 0.928 | 0.950 | | 0.465 | 0.428 | 0.400 | 0.435 |
| Position asymmetry | 0.326 | 0.381 | 0.318 | 0.355 | 0.392 | 0.321 | 0.320 | 0.344 | | 0.953 | 0.937 | 0.979 |
| Periodic Asymmetry Index | 0.299 | 0.363 | 0.283 | 0.369 | 0.266 | 0.256 | 0.276 | 0.292 | 0.531 | | 0.924 | 0.912 |
| Average Mutual Information | 0.225 | 0.267 | 0.216 | 0.247 | 0.314 | 0.217 | 0.228 | 0.251 | 0.873 | 0.469 | | 0.911 |
| Fourier spectrum | 0.381 | 0.455 | 0.373 | 0.432 | 0.414 | 0.363 | 0.370 | 0.401 | 0.914 | 0.713 | 0.726 | |

CU    Codon Usage
HU    Hexamer Usage
CPre  Codon Preference
AAU   Amino Acid Usage
CPro  Codon Prototype
MM-1  Markov Model, k = 1
MM-2  Markov Model, k = 2
MM-3  Markov Model, k = 3
PA    Position Assymetry
PAI   Periodic Assymetry Index
AMI   Average Mutual Information
FOU   Fourier Spectrum

strongly correlated (Fickett & Tung, 1992). Table 4.6 shows the correlation between the scores of the coding statistics reviewed here in the set of human exon and intron sequences: as can be seen, the coding statistics are strongly correlated. The only two statistics truly uncorrelated (as expected) are codon preference and amino acid usage in exonic sequences. It appears, therefore, that some combination of just two statistics, one measuring correlation between positions within a codon, and the other measuring dependence between codons along the query sequence could produce the most discriminant output.

# REFERENCES

Borodovsky M. & McIninch J. (1993) *Comp. Chem.* **17**, 123–130.
Burge C. & Karlin S. (1997) *J. Mol. Biol.* **268**, 78–94.
Burset M. & Guigó R. (1996) *Genomics* **34**, 353–357.
Claverie J.-M. (1997) *Hum. Mol. Genet.* **6**, 1735–1744.
Claverie J.-M., Sauvaget I. & Bougueleret L. (1990) *Meth. Enzymol.* **183**, 237–252.
Dong S. & Searls D.B. (1994) *Genomics* **23**, 540–551.
Fichant G. & Gautier C. (1987) *Nucl. Acids Res.* **4**, 287–295.
Fickett J.W. (1982) *Nucl. Acids Res.* **10**, 5303–5318.
Fickett J.W. (1996) *Trends Genet.* **12**, 316–320.
Fickett J.W. & Tung C.S. (1992) *Nucl. Acids Res.* **20**, 6441–6450.
Fields C.A. & Soderlund C.A. (1990) Gm: a practical tool for automating DNA sequence analysis. Computer Applications to the Bioscience, 263–270.
Gelfand M.S. (1995) *J. Comput. Biol.* **1**, 87–115.
Gelfand M.S. & Roytberg M.A. (1993) Prediction of the exon-intron structure by a dynamic programming approach. *BioSystems*, **30**, 173–182.
Gelfand M.S., Mironov A.A. & Pevzner P.A. (1996) Gene recognition via spliced alignment, *Proceedings National Academy Sciences USA* **93**, 9061–9066.
Grantham R., Gautier C., Gouy M., Mercier R. & Pave A. (1980) *Nucl. Acids Res.* **8**, 49–62.
Gribskov M., Devereux J. & Burgess R.B. (1984) *Nucl. Acids Res.* **12**, 539–549.
Grosse I., Herzel H., Buldyrev S.V. & Stanley H.E. (1998) unpublished.
Guigó R. (1997a) *J. Mol. Med.* **75**, 389–393.
Guigó R. (1997b) *Comput. Chem.* **21**, 215–222.
Guigó R., Knudsen S. & Drake N. (1992) Prediction of Gene Structure. *J. Mol. Med.* **226**, 141–157.
Gutiérrez G., Oliver J. & Marín A. (1994) *J. Theor. Biol.* **167**, 413–414.
Henderson J., Salzberg S. & Fassman K.H. (1997) *J. Comput. Biol.* **4**, 127–141.
Herzel H. & Grosse I. (1995) *Physica A* **216**, 518–542.
Hutchinson G.B. & Hayden M.R. (1992) The prediction of exons through an analysis of spliceable open reading frames. *Nucl. Acids Res.* **20**, 3453–3462.
Ikemura T. (1985) *Mol. Biol. Evol.* **2**, 13–34.
Konopka A.K. (1994) Towards mapping functional domains and indiscriminantly sequenced nucleic acid: a computational approach. In *Structure and Methods: VI.*

*Human Genome Initiative and DNA Recombination*, (eds.) R.H. Sarma and Sarma, M.H. Adenine Press, Guiderland NY, pp 113–125.

Kulp D., Haussler D., Reese M. & Eeckman F.H. (1996) In *Intelligent Systems for Molecular Biology*, D.J. States, P. Agarwal, T. Gaasterland, L. Hunter and R. Smith (eds.) AAAI Press, Menlo Park, pp. 134–142.

Li W. (1997) *Comp. Chem.* 21, 257–271.

Li W., Marr T.G. & Kaneko K (1994) *Physica D* 75, 392–416.

Marin A., Bertranpetit J., Oliver J.L. & Medina J.R. (1989) Variation in G+C content and codon choices: differences among synonymous codon group in vertebrate genes. *Nucl. Acids Res.* 15, 6181–6189.

McCaldon P. & Argos P. (1988) *Prot. Struct. Funct. Genet.* 4, 99–122.

Milanesi L., Kolchanov N.A. & Rogozin I.B. et al (1993) A computing tool for protein-coding regions prediction in nucleotide sequences, Proceedings 2nd International Conference on Bioinformatics, Supercomputing and Complex Genome Analysis (July 1992, St. Petersburg, Florida), 573–587, H.A. Lim, J.W. Fickett, C.R. Cantor & R.J. Robbins eds., Singapore, World Scientific.

Mural R.J., Mann R.C. & Uberbacher E.C. (1991) Pattern recognition in DNA sequences: The intron-exon junction problem. C.C Cantor & H.A. Lim (eds) *Proceedings of the First International Conference on Electrophoresis, Supercomputing and the Human Genome*. World Scientific Co. Singapore, pp. 164–172.

Shannon C.E. (1948) *Bell Syst. Bell Syst. Techn. J.* 27, 379–423.

Shepherd J.C. (1981) *Proc. Natl. Acad. Sci. USA* 78, 1596–1600.

Silverman B.D. & Linsker R. (1986) *J. Theor. Biol.* 118, 295–300.

Snyder E.E. & Stormo G.D. (1993) Identification of coding regions in genomic DNA sequences: an application of dynamic programming and neural networks. *Nucl. Acids Res.* 21, 607–613.

Solovyev V.V., Salamov A.A. & Lawrence C.B. (1994) *Nucl. Acids Res.* 22, 5156–5163.

Staden R. (1984) *Nucl. Acids Res.* 12, 551–567.

Staden R. & McLachlan A. (1982) *Nucl. Acids Res.* 10, 141–156.

Thomas A. & Skolnick M.H. (1994) A probabilistic model for detecting coding regions in DNA sequences. *IMA Journal of Mathematical Applications to Medicine and Biology*, 11, 149–160.

Tiwari S., Ramachandran S., Bhattacharya A., Bhattacharya S. & Ramaswamy R. (1997) *Comput. Appl. Biosci.* 13, 263–270.

Tsonis A.A., Elsner J.B. & Tsonis P.A. (1991) *J. Theor. Biol.* 151, 323.

Uberbacher E.C. & Mural R.J. (1991) *Proc. Natl. Acad. Sci. USA* 88, 11261–11265.

Xu Y., Mural R.J. & Uberbacher E.C. (1994) Constructing gene models from accurately predicted exons. *Computer Applications in the Biosciences*, 11, 117–124.

Zhang M.Q. (1997) Identification of protein coding regions in the human genome based on quadratic discriminant analysi. *Proceedings of National Academy of Science* (USA), 94, 559–564.

# 5 The Properties of Amino Acids in Sequences

*William R. Taylor*

Division of Mathematical Biology, National Institute for Medical Research,
The Ridgeway, Mill Hill, London, UK

## 5.1 INTRODUCTION

To a first approximation, all the cellular machinery that constitutes life is based on proteins. This incredibly diverse range of functions is achieved through the equally varied structures that proteins can adopt. Further dissection of this complexity, however, reveals a bifurcation – on one side complex but on the other simple. The simplicity is that proteins are linear polymers composed of only 20 amino acids while the complexity is retained in the virtually infinite orderings of these 20 units along the polypeptide chain. One of the most fundamental and fascinating aspects of protein science is the interplay of these levels in producing a folded functional protein, often epitomized by the structure prediction problem: 'given a protein sequence, what is the three-dimensional structure?' At first sight, the problem appears to be tractable: the protein sequence is known exactly, as is the chemical structure of the main chain and each side-chain – yet such is the complexity of the possible interactions within the chain (also involving solvent), that computation of a folded structure (at least of realistic size) has, so far, proved impossible.

With the failure of (or some would say hiatus in) the fundamental *ab initio* approach to the prediction of protein structure and function, other less direct approaches have been pursued. These rely mainly on empirical inference of the kind: 'given we have seen this type of residue buried in all known structures, it should also be buried in the structure of all proteins whose structures we have yet to see'. Compared with the *ab initio* approach, the empirical approach is inherently weak, as the protein of current interest might be the exception that breaks the hitherto perfect rule. Such inferences are especially fraught where amino acids are involved as they can change both their properties and the relative importance of different interactions,

*Genetics Databases*
ISBN 0-12-101625-0

depending on their situation in the protein. The former effect might be simply a change in ionization state allowing a normally charged residue to be buried (away from solvent). The latter effect derives from the multifunctional nature of most amino acid side-groups, which allows one function (such as donating a hydrogen-bond) to be used in one situation in the structure while another function (such as accepting a hydrogen-bond) might be important in another context. Which function is used in which place is only apparent in the folded structure, which itself depends on how the amino acids deploy their different functions in the first place. Breaking this circularity would be a step towards predicting structure.

## 5.2    PROPERTIES OF AMINO ACIDS

Before considering the subtle interactions of amino acids in a complex protein structure it is best first, to consider their properties individually.

### 5.2.1    Physico-chemical properties

#### 5.2.1.1    Physical properties

The basic properties of the amino acids side chains can be typified by the number of atoms (size), the number of polar[1] atoms (N, O, and to a lesser extent S) that they contain and the degree to which the atoms are constrained (flexibility).

Table 5.1 lists the most basic properties of the amino acids as numbers of atoms and bonds. These simple counts are sufficient to identify each amino acid as unique and even without distinguishing polar atom type, only two pairs of acids have identical properties. Adopting this simplification gives three coordinates that allows the acids to be 'plotted' in space (Fig. 5.1). This representation shows the large size range from Gly to Trp and also how the nominally polar Lys tends to associate with Met (both have long flexible side-chains) using the third dimension to avoid 'dragging' the other polar and non-polar amino acids into close proximity.

---

[1] The term *polar* is used loosely throughout this chapter to refer to atoms that are sufficiently electronegative relative to carbon, and especially hydrogen, to induce electrostatic charge polarization across the bonds that they form. This polarity gives rise to hydrogen-bonding which is critical in the structure and function of proteins. A detailed consideration of the topic can be found in Kyte (1995) (Chapter 2, on the electronic structure of the amino acids).

**Table 5.1** Physical properties of the amino acids. The 20 amino acids are tabulated (with their abbreviations) showing their number of non-hydrogen atoms (beyond the $\alpha$-carbon), the number of which are polar (non-carbon) and the number of bonds (including the $\alpha$-$\beta$ carbon bond) which can be rotated to move the position of non-hydrogen atoms. The latter property corresponds to flexibility.

| Amino acid | Single-letter code | Three-letter code | All atoms | Polar N, O, S | Rotate bonds |
|---|---|---|---|---|---|
| Aspartic acid | D | Asp | 4 | 0,2,0 | 2 |
| Serine | S | Ser | 2 | 0,1,0 | 1 |
| Threonine | T | Thr | 3 | 0,1,0 | 2 |
| Glycine | G | Gly | 0 | 0,0,0 | 0 |
| Proline | P | Pro | 3 | 0,0,0 | 0 |
| Cysteine | C | Cys | 2 | 0,0,1 | 1 |
| Alanine | A | Ala | 1 | 0,0,0 | 0 |
| Valine | V | Val | 3 | 0,0,0 | 2 |
| Isoleucine | I | Ile | 4 | 0,0,0 | 2 |
| Leucine | L | Leu | 4 | 0,0,0 | 3 |
| Methionine | M | Met | 4 | 0,0,1 | 3 |
| Phenylalanine | F | Phe | 7 | 0,0,0 | 2 |
| Tyrosine | Y | Tyr | 8 | 0,1,0 | 2 |
| Tryptophan | W | Trp | 10 | 1,0,0 | 2 |
| Histidine | H | His | 6 | 2,0,0 | 2 |
| Arginine | R | Arg | 6 | 3,0,0 | 4 |
| Lysine | K | Lys | 5 | 1,0,0 | 4 |
| Asparagine | N | Asn | 4 | 1,1,0 | 2 |
| Glutamine | Q | Gln | 5 | 1,1,0 | 3 |
| Glutamic acid | E | Glu | 5 | 0,2,0 | 3 |

## 5.2.1.2 Chemical properties

The most hydrophobic residues (those composed entirely of carbon) have, on the energy scales of proteins, effectively no chemistry. Binding specificity and enzymic activity reside in the variety of chemical function associated with the polar atoms in the more hydrophilic residues. What was classed above simply into nitrogen, oxygen and sulphur containing groups, in chemical terms, become acids (Asp, Glu), bases (Arg, Lys), hydroxyl (Ser, Thr, Tyr) amide (Asn, Gln), imidazol (His) and sulphydryl (Cys). From a structural viewpoint, the differences in these functions are manifest mainly in their differing propensity to form hydrogen bonds or salt-bridges (between the acids and bases). Most polar atoms can both accept a hydrogen bond (onto their electronegative

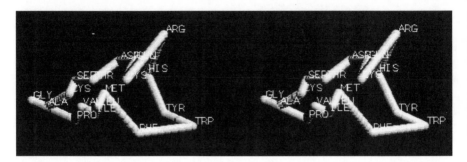

**Figure 5.1** Physical properties of the amino acids (defined in Table 5.1) are plotted in space (as a stereo pair). The grey lines connect amino acids so as to avoid crossing lines and corresponds well to intuitive notions of similarity. Since only the number of polar atoms are plotted, the pairs Glu/Gln and Asp/Asn are unresolved. In addition, before plotting, the count of all polar atoms was doubled, with the exception of sulphur.

orbitals) or donate a hydrogen bond (through an attached hydrogen). Details of these bonding options can be found in Kyte (1995) and the exhaustive (if now slightly dated) survey by Baker & Hubbard (1984).

The activity of the various chemical functions in catalysis are so many and varied that any review is impossible here. A rough guide, however, is that acid and basic groups are often used to polarize bonds so that they become 'activated' ready to perform a chemical reaction. Water is often activated in this way and can then proceed to break other bonds. The wide class of enzymes referred to collectively as hydrolases use this mechanism. Most standard biochemical textbooks contain a comprehensive survey of enzymatic catalysis, however, of especial relevance is the recent work of Kyte (1995) which approaches the topic from a more chemical perspective.

Few of the chemical groups on amino acid side-chains change their ionization state within the range of physiological environments, with the exception of His. This property often gives His a unique role in catalysis, perhaps best demonstrated in the His-Asp-Ser catalytic triad, classically found in the serine proteases but also independently evolved in other enzymes (Fischer *et al.*, 1994). In the active site of the serine proteases, the hydroxyl (OH) group is polarized (activated) by a positively charged His which is stabilized from the other side by the (negative) acid Asp. On reaction, the Ser donates its hydrogen (leaving O⁻) but then regains it from the His. The resulting uncharged His, however, can still hydrogen-bond to the Asp and eventually 'picks-up' another proton to reconstitute the original state of the triad.

Where the amino acids themselves do not have sufficient electronegativity or polarizability in their limited range of atoms to perform a reaction, they often 'recruit' other cofactors, in particular, nucleotides and metals. The acids involved in this binding are, again, His and Cys (especially for iron, zinc and copper), with Met also being found binding copper (Yamashita *et al.*, 1990).

### 5.2.1.3   Post-translational modification

Some acids occasionally become modified after translation, the most common of which include phosphorylation, the methylation of Lys, the addition of another acid group to Glu (γ-carboxy glutamic acid), the addition of a hydroxyl group to Pro (and Tyr) and the addition of sugars to Asn – for a more exhaustive list, including modifications to the main-chain, see Kyte (1995), Chapter 2, and Martin (1997). With the exception of phosphorylation and gly-cosylation, these events are fairly uncommon and as they are enzymic modifi-cations to an already folded (or partly folded) protein (Martin, 1997), they are not major factors in determining the fold of the protein and, as such, they are not of any great general structural importance. Given, also that they are not easy to predict from just the protein sequence[2], it is difficult to consider them in sequence analysis.

The most common modification of the protein after translation is the cross-linking of two cysteines (Cys) to form a cystine (or disulphide bridge). This modification is associated with proteins that are exported from the cell into the external oxidizing environment. As with the above modifications, disulphide formation appears to stabilize the final structure and is not a determining fac-tor in which fold the protein chain adopts. This can be deduced from the action of protein-disulphide isomerase which actively aids the rearrangement of disul-phide bonds until the correct pairs can be found. Disulphide-bridge formation is largely suppressed inside the cell by the reducing activity of glutathione (and its reductase). This differing behaviour of Cys results in different criteria for the assessment of the structural and functional role of Cys depending on whether the protein in which it occurs is known to be exported from the cell. A conserved pair of Cys in an extracellular protein might reasonably be inter-preted as likely candidates to form a disulphide bond whereas if the protein was intracellular, metal binding might be a more probable cause of conserva-tion.

---

[2] Although potential phosphorylation and Asn-linked glycosylation sites are easy to predict, it can-not be determined from the sequence whether modification will occur at all of them.

## 5.2.2    Steric properties

### 5.2.2.1    Glycine and proline

The steric properties of amino acids can greatly influence the local structures that they adopt (or 'favour'). This is most clearly seen in Pro, which has its side-chain back-bonded to the amide nitrogen of the main-chain, resulting in reduced torsional freedom of the resudue (with the main-chain $\phi$ angle fixed). The conformation of the backbone is locked into a turn, which, combined with the loss of hydrogen-bonding potential on the nitrogen, results in Pro being associated with surface loops. This sterically induced role is at variance with the pure carbon (and therefore hydrophobic) composition of its side-chain, making Pro the exception in many classification schemes.

Gly similarly defies classification – since, having no side-chain, there is little to classify it with. In contrast to Pro, the lack of side-chain gives increased torsional freedom around both the rotatable main-chain bonds ($\phi$ and $\psi$ angles) but, paradoxically, this leads to the same structural preference as Pro. Like Pro, Gly is also associated with surface loops and turns where its unhindered rotational freedom can result in increased loop flexibility (sometimes desirable for facilitating conformational change) and also in the capacity to allow tight turns in the main-chain. Since Gly has no side-chain, it also has no polar atoms and so can be found buried in the hydrophobic core of the protein. This dual nature often makes it difficult to predict the structural significance of Gly when seen in a sequence.

### 5.2.2.2    Secondary structure preferences

While Gly and especially Pro have a propensity to avoid secondary structure, the other amino acids have, to varying degrees, a preference to adopt a particular secondary structure. The $\alpha$-helix conformation favours amino acids that do not have too much bulk, or polar atoms, close to the $\alpha$-carbon. This mitigates against side-chains that branch at the $\beta$-carbon (Val, Ile, Thr) and the small hydrophilic residues (Ser, Asp, Asn and Thr again). Of the remainder, the hydrophilic acids Glu, Gln, Arg and, to a lesser extent, Lys are favoured and of the hydrophobic residues, Ala, Leu, Met and Phe are especially associated with the $\alpha$-helix. The converse is true of the $\beta$-sheet secondary structure in which residues that branch at the $\beta$-carbon pack together well along a strand, often stacking together in a chevron arrangement (<–<–<, where each angle represents the $\beta$-carbon with its two attached $\gamma$-carbons, with each residue occupying alternate positions along a $\beta$-strand). In the $\beta$-sheet, the hydrogen-bonds are better shielded from the side-chain, allowing the occurrence of polar atoms near the main chain without disrupting the secondary structure.

The above analysis is made in the context of the hydrogen-bonded secondary structure. However, the acids also have an intrinsic propensity to adopt any particular local substructure (characterized by the back-bone φ, ψ angles). To examine this, it is necessary to separate the acids from the context of the secondary structure. Secondary structure is preferentially found in the core of the protein as it is here that the polar backbone amide and carbonyl functions must be locked away (in the form of hydrogen bonds) from the hydrophobic environment. Because hydrophobic amino acids are also associated with the core, they will appear to have a 'false' propensity to adopt secondary structure. To allow for this context-dependence complication Swindells *et al.* (1995) analysed the torsion angles of residues that were not in secondary structure, thus 'factoring-out' the effect of hydrogen-bonding. From this analysis, the acids were found largely to retain the preferences outlined above, with the exception of Thr, Asp and Ser which had an intrinsic preference to prefer the α-conformation.

### 5.2.2.3 Core packing

From the viewpoint of the stability of globular proteins, the ability to form a well-packed hydrophobic core is critical. This is reflected in the subtle variation of shape and size seen in the hydrophobic amino acids, with Val, Ile and Leu differing only by the addition and rearrangement of a methyl group.

Some of the longer chain hydrophilic acids (Arg, Lys, Glu, Gln and Tyr) are often associated with the edge of the hydrophobic core, being 'rooted' in the core but 'growing' out towards the solvent where their polar atoms can form hydrogen-bonds with other polar residues or solvent. This effect is seen even at the sequence level, in which a position that is otherwise hydrophobic across family members will be substituted by a long-chain acid.

## 5.3 EMPIRICALLY DERIVED AMINO ACID RELATIONSHIPS

The properties listed in Table 5.1 are largely physical in nature and the derived portrayal of the relationships of the acids in Fig. 5.1 contains little reflection of the chemistry of the groups – for example, whether a residue is acidic, basic, aromatic or aliphatic. Much effort has, in the past, been invested in such analysis and categorization: for example, Sneath, 1966) and has provided the basis for measures of amino acid similarity (McLachlan, 1971) (see Taylor, 1986 for a review). However, irrespective of the quality of the physico-chemical analysis, unless the properties are discriminated in the same way within the protein structure (by the protein itself), they will have little predictive value. This problem can be seen more simply by reference to Table 5.1, where the scale of size (by atom count) ranges from zero to 10, while the scale of polarity (by number of polar atoms) ranges from zero to three. This, of course, does not imply that size is three times more important to protein structure and function than polarity, and the same applies to any physico-chemical measure.

## 5.3.1 Sequence substitution matrices

By observing the degree to which the amino acids exchange over the course of evolution it is possible to extract a rough idea of the relative importance of properties – for example, by counting the frequencies of substitution at equivalent sites in homologous proteins. This analysis, however, is indirect and the resulting relationships among the acids can only be interpreted afterwards in terms of physico-chemical properties. Such analysis also suffers from the problem that averages must be made over a wide variety of proteins; and, as was seen above, different properties will acquire different significance depending on the environment in which the protein functions (for example; extracellular/intracellular or thermophilic/mesophilic). This problem also extends to the 'micro-environments' within a protein structure (for example; buried/exposed or secondary structure state).

The resulting matrices of pairwise relationships between amino acids are typified by the 'classic' PAM series calculated by Dayhoff and coworkers (Dayhoff *et al.*, 1978), which has been revised and recalculated in different ways a number of times (Jones *et al.*, 1992; Henikoff & Henikoff, 1992; Gonnet *et al.*, 1992). Such matrices are directly useful, principally for the alignment of sequences, independently of their physical meaning: however, it is of benefit to their fuller understanding to try to gain some insight into what they represent and what aspects are emphasized in each variant. It is difficult to gain this understanding simply from looking at the 20 × 20 field of numbers that constitutes the data, but by extracting a spatial representation of the matrix of numbers, some structure can be seen – in much the same way as Fig. 5.1 is easier to appreciate than the corresponding set of numbers (Table 5.1).

While Table 5.1 could be directly represented in three dimensions, an amino acid similarity matrix bears little relationship to a three-dimensional object. This problem, however, can be solved by a technique sometimes referred to as multidimensional scaling, in which a set of relationships are converted to pairwise distances and the distances then projected (or embedded) into a Euclidean space (French & Robson, 1983). Unfortunately, there is no guarantee that this space will have three (or fewer) dimensions (or even that it will be Euclidean) and some experimentation is often necessary to find a suitable formulation (Taylor & Jones, 1993). Low-dimensional results can, nonetheless be obtained (without significant loss of information) that lead to representations of the amino acids that can easily be interpreted in terms of the basic properties discussed above. A projection of the Dayhoff $PAM_{250}$ matrix as a two-dimensional figure is shown in Fig. 5.2 (taken from (French & Robson, 1983) and oriented for easy comparison with Fig. 5.1). The disposition of the amino acids in the projection is not dissimilar to that in the much more simply derived Fig. 5.1. The major axis clearly corresponds to size (Gly–Trp) while the second axis corresponds to polarity.

By generating their spatial forms, matrices can be compared. In Fig. 5.3, the Dayhoff $PAM_{250}$ matrix and the more recent matrix resulting from a similar calculation by Jones *et al.* (1992) (the JTT matrix) are projected in three dimensions and superposed (as if they were protein structures) to reveal differences in the location of the amino acids. One of the visually unsatisfactory aspects of such plots is the large range of similarity. Some acids lie very close together while others, such as Trp, are distant. Using least-squares refinement, the close relationships between the acids can be maintained at a reasonable distance while keeping the overall structure (see Taylor & Jones, 1993, for further details).

The distance matrix from which the spatial form is derived can also be of direct use in sequence analysis where a true metric of sequence similarity is

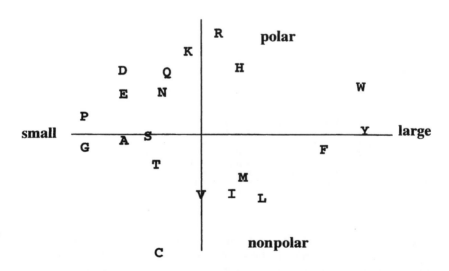

**Figure 5.2** Two-dimensional representation of the Dayhoff matrix (adapted from French & Robson, 1993) showing the dominant axes of size and polarity. This result effectively gives a scale to the dimensions of Fig. 5.1 (which were otherwise arbitrary). The relative positions of the acids are reasonably similar.

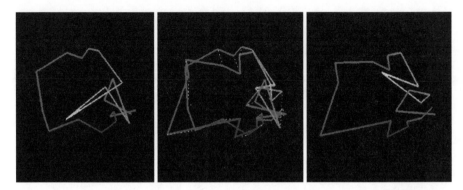

**Figure 5.3**   Three-dimensional comparison of the PAM$_{250}$ and JTT matrices. The two matrices were projected into 3D and amino acids connected and coloured as specified in Table 5.3 (starting at Asp). Left panel, JTT matrix and right panel PAM$_{250}$ matrix. The centre panel shows the least-squares superposition of the structures on equivalent acids (connected by dashed lines). Cys was not included in the superposition (or plot) as it occupies quite distinct positions in the two structures lying below the ring of acids in the PAM$_{250}$ matrix and above the ring (close to Trp) in the JTT matrix. The most significant difference between the structures otherwise is in the location of Gly which moves away from the polar group in the JTT matrix.

```
A
20.4   C
 8.7 25.0   D
 8.1 24.8  2.6   E
24.3 25.7 28.7 27.5   F
 6.8 21.1  8.6  9.1 27.0   G
11.7 24.6 10.8 10.1 23.9 15.1   H
12.8 21.9 18.5 17.2 16.9 17.2 16.7   I
10.7 25.3 10.3  9.6 26.2 13.4 10.3 16.7   K
17.3 26.1 22.1 20.5 14.0 21.0 19.1  8.8 19.1   L
14.1 25.6 18.9 17.2 18.3 18.4 16.6  7.5 14.7  6.7   M
 7.2 23.4  5.8  5.5 25.0  9.7  6.9 15.8  7.2 19.1 15.9   N
 7.0 21.9 11.0 10.5 26.7 10.4 12.5 16.1 11.2 19.7 16.5  9.9   P
 9.3 25.8  6.2  5.2 27.2 11.9  7.1 17.1  7.5 19.6 16.0  5.3 10.1   Q
14.3 26.2 14.9 14.2 25.4 17.4 10.5 18.1  7.1 19.4 16.0 10.8 13.4 10.9   R
 5.7 20.0 10.6 10.3 24.0  9.2 11.0 14.1  9.9 18.5 15.5  7.2  8.6 10.6 11.6   S
 3.6 20.6  9.7  8.8 23.3  8.7 11.4 11.6  9.7 16.5 13.1  7.2  8.3  9.9 13.1  5.1   T
10.5 21.4 16.8 15.6 19.2 14.9 16.3  3.6 15.6 10.1  7.7 14.6 14.0 15.7 17.6 12.7  9.9   V
32.1 34.4 34.0 33.5 24.2 33.3 29.6 30.6 28.7 26.7 29.1 30.2 32.0 31.8 24.5 29.3 31.1 31.7   W
23.0 22.3 26.8 25.9  7.9 25.7 21.2 18.3 25.0 17.7 20.4 22.8 25.6 25.5 24.2 22.1 22.2 20.3 25.0   Y
```

**Figure 5.4**   Distance matrix derived from the PAM$_{250}$ matrix. The diagonal (which is all zero) is not shown. The PAM$_{250}$ matrix from which the distance matrix is derived had 10 added to all entries to remove negative numbers. The ranked eigen values from the projection of the matrix (Fig. 5.3), expressed as a percentage of the total, were 45.7, 20.8, 14.8, 8.5, 3.6, 2.2, 1.4, 1.1, 0.4, 0.4, 0.2 and 0.1, indicating that most of the information (81.3%) is contained in the first three dimensions.

required (Vingron & von Haeseler, 1997). Unlike a similarity matrix, a distance matrix has the property that two identical sequences always have the same alignment score (zero) irrespective of their length or composition. The distance matrix corresponding to the Dayhoff PAM$_{250}$ matrix is shown in Fig. 5.4.

## 5.3.2    Environment-specific matrices

### 5.3.2.1    Sequence/sequence applications

The problem of averaging over distinct types of protein and distinct substructures within proteins can be overcome by compiling separate amino acid exchange matrices for each state. For application in sequence analysis, this approach is useful only if the states can be predicted beforehand, so that the correct matrix can be applied at the correct point. Sometimes, such prediction is easy as information, independent of the sequence, might be available to guide in the choice: for example, it is often known whether a protein is intra- or extra-cellular. Such information can also be obtained from the prediction of transmembrane segments in proteins: identifying one part of the sequence as external and the other internal.

The prediction of transmembrane segments themselves is relatively simple from sequence data and an amino acid exchange matrix has been derived for this specific environment (Jones *et al.*, 1994). As might be expected, it is quite distinct from the standard matrices derived from (aqueous) globular proteins and has been applied selectively during the alignment of membrane associated sequences (Taylor & Jones, unpublished data).

Amino acid exchange matrices have been calculated for each secondary structure state ($\alpha$-helix, $\beta$-sheet and other) and reveal quite distinct preferences. So distinct, indeed, that the analysis of observed exchanges can provide a good guide to the secondary structure state itself (Mehta *et al.*, 1995). This approach can be further abstracted to predict simultaneously the secondary structure state and refine the phylogenetic interrelationship among the sequences in a multiple alignment, using a hidden Markov model (Thorne *et al.*, 1996).

### 5.3.2.2    Sequence/structure applications

In the application of structure-specific matrices to sequence alignment, half of the uncertainty can be removed if the structure of one of the sequences is known. The alignment of a sequence with a known structure is variously known as '3D/1D alignment' or 'threading' (with the latter approach making more use of direct three-dimensional interactions). In this situation the structural environment of one side of the alignment problem is known and can be used to select the substructure specific matrix used to evaluate the match. Simple applications began by discriminating only secondary structure state and degree of burial (Bowie *et al.*, 1991) which was later elaborated into finer distinctions (Rice & Eisenberg, 1997). In parallel, Blundell and coworkers identified several structurally distinct environments and derived substitution matrices for each (Overington *et al.*, 1990, 1992).

## 5.4    RELATIONSHIP TO THE GENETIC CODE

Having reviewed the properties of the 20 biological amino acids, one might ask why just 20 (or even why as many as 20) and why these particular 20 (who's idea was it to have Trp anyway?), indeed, why amino acids at all? The last question has no obvious answer except that amino acids were probably abundant in the 'primordial soup' and the subsidiary question of 'why just L-amino acids?' has been answered by the observation that, in coexistence, the two enantiomers would be in competition for basic resources and, given the time periods involved, one system must win. While little more can be said about these most ancient beginnings, some slightly more concrete speculations can be made about later developments, based on the structure of the genetic code.

### 5.4.1    The origin of the acids

It was first postulated by Crick (1968) that there were originally only a few acids that were gradually augmented to through the extension of the genetic code to its current full triplet-based coding system. It can then be speculated that the acids with the most synonymous codons represent the original repertoire which have since had codons progressively reallocated. This logic suggests that there were originally three acids Arg, Ser and Leu as each of these currently has six codons. From a physico-chemical viewpoint, this is not an unreasonable selection as each are widely spaced in physico-chemical space. Arg is large and positive, Leu is medium sized and hydrophobic while Ser is small (and slightly negative). If plotted on the size–polarity plane, they form a well-separated triangle. (See Fig. 5.2). All the codons for Leu have uracil (U) in the middle position, while Ser codons mostly have cytosine (C) and Arg all have guanine (G) in the middle position. It is possible that adenosine (A) in the middle might have been a STOP codon, but whether these ancient messages were sufficiently long to justify a STOP is debatable. For discussion and elaboration of some of these ideas, see Gibson & Lamond (1990) and also Lamond & Gibson (1990).

While it is possible to imagine simple proteins formed from just these three amino acids, Gibson & Lamond (1990) have suggested that the proteins originally encoded were probably integral membrane proteins, for which even Leu and Ser alone might suffice. With the evolution of globular proteins, the need for better core packing and, in particular, the need to have a β-sheet in the core probably provided the drive to recruit the first codon position to distinguish either Ile or Val from Leu. This would allow the preference for α-helix and β-sheet to be encoded in the sequence – a critical capacity since most of the very ancient enzymes have an alternating β/α structure. In the code, the rougher distinction of just purine over pyrimidine in the first position would be sufficient to initially distinguish a β-branched acid which was later elaborated using either A or G to further distinguish Val from Ile. The recruitment of Met and Phe to the hydrophobic collection made use of the third position and (with one codon each) would appear to be the most recent addition.

The possible evolution from an original Ser (with C in the middle position) is less amenable to analysis (or speculation) since each of the present-day acids with C in the second position (Pro, Thr and Ala) are uniquely specified by their first codon position. All, however, are very similar to Ser, with Ala and The being just an oxygen and methyl substitution away, while Pro is also small and found in surface turns. It is easy to imagine that a change in the code (which at some point would have arbitrarily changed half of the Ser in proteins to Thr) could have been permissible.

The Arg-dominated column in the code (with G in the second position) appears to have had arbitrary encroachments from fairly unrelated acids, including Gly, Ser, Cys and Trp (although the last as some affinity to Arg). Similarly, the column with A in the second position is evenly populated by the remaining mainly large hydrophilic residues, giving no hint of its possible evolutionary development.

## 5.4.2    Redundancy in the genetic code

Whether as a result of its evolutionary history (as speculated above), or through an evolved design, the genetic code is tolerant to random mutation in much the same way as signal engineers design their communication codes to be tolerant of 'noise-on-the-line'. The approach used in signal transmission is to have a code in which changing one bit (say, a dot to a dash, in the Morse code) makes a minimal change in the encoded information (referred to as a Grey coding). The four nucleotides used in the genetic code require two bits to represent them, which, combined with the three codon positions gives six bits (corresponding to the 64 codons). These can then be arranged in a (circular) order such that only one bit changes between adjacent positions (see Taylor, 1989, for a familiar example). Mapping the four nucleotides onto this circle, however, is not unique since every ordering of the nucleotides gives rise to a different circle of codons. One such ordering was found by Swanson (1984) to correspond well to the major physicochemical properties of size and polarity and, as can be seen from Table 5.2, many of the other possibilities do so also.

**Table 5.2**  Amino acid orderings from Grey-coded codons. The binary numbers 0–63 were arranged so that only one bit changes between adjacent numbers (Grey coded) each pair of digits can be taken to represent a nucleotide (e.g. U = 00, C = 10, A = 01 and G = 11) and all 24 possible assignments are applied (key) giving ordered codons in which only one base changes between adjacent entries. Translating the codons to their amino acid (where X = STOP) gives a list of 64 acids. This list is simplified by unifying identical adjacent acids giving the string shown next to each key. The shorter strings (shown with bold keys) are better, as point mutations will be less likely to change the encoded amino acid. Two sets of data are shown in which (a) the first codon position is assigned to the most slowly changing index and (b) the second codon position changes most slowly. Because the encoding is circular, the first and last acids are also adjacent.

| Key | Amino acid order |
| --- | --- |
| **(a) Position 1 slow** | |
| UCAG | VAGEDNKRSTIMIFLSCWXYHQRPL |
| UCGA | IMTNKRSGEDAVFLSYXWXCRQHPL |
| UGCA | IMISRSNKNTPHQHRLFLFCXWCYXYSADEDGV |
| GUCA | RSRSIMKNKNTPHQHQLRGVEDEDASYXYXLFLFCXCW |
| UACG | VDEDGAPRHQHLFLFYXYCXWCSTSRSNKNIMI |
| UAGC | LQHQHPRGADEDEVFLFLXYXYSWCXCSRSRTNKNKMI |
| UGAC | LRPQHQHNKNKTSRSRIMIFLFLXCWCSXYXYDEDEAGV |
| GUAC | RLPQHQKNKTMIRSRGVAEDEXYXSLFLXCW |
| AUCG | EDEDVGAPRLHQHQKNKNIMIRSRSTSCWCXLFLFYXYX |
| AUGC | QHQLPRGAVEDEKNKMITRSRXCWSLFLXYX |
| AGUC | QHRPLFSXWCYXKNSRTIMIVAGDE |
| **GAUC** | RHQPLFSXYCXWGDEAVMITKNS |
| CUAG | AVGEDNKRSIMITPLRQHYXWCFLS |
| CUGA | TMINKRSGEDVAPLHQRCXWXYFLS |
| CGUA | TSRSNKNIMIFLFYXYCWXCSPRHQHLVDEDGA |
| GCUA | RSRSTKNKNIMLFLFYXYXSCXCWGAEDEDVLHQHQP |
| CAUG | ADEDGVFLFCWXCYXYSPHQHRLIMISRSNKNT |
| CAGU | SXYXYFLFLWCXCGVDEDEAPQHQHLRSRSRMINKNKT |
| CGAU | SXCWCFLFLXYXYNKNKIMISRSRTPRLQHQHDEDEVGA |
| GCAU | WCXSLFLXYXKNKIMTRSRGAVEDEQHQLPR |
| ACUG | EDEDAGVLFLFCWCXSYXYXKNKNTRSRSIMILRPHQHQ |
| ACGU | XYXSLFLWCXGVAEDEKNKTIMRSRLPQHQ |
| AGCU | XYCWXLFSPLRHQKNSRIMITAVGDE |
| GACU | WXCYXLFSPLQHRGDEVATIMKNSR |

**(b) Position 2 slow**

| | |
|---|---|
| UCAG | CWXRGRSNKEDHQXYFLVIMITAPS |
| **UCGA** | YXQHNKEDGRSRWXCFLIMVATPS |
| UGCA | YXYDEDNKNHQHPTASFLFVIMILRSRSGCXWC |
| GUCA | EDEDYXYXKNKNHQHQPTSAGCXCWRSRSRLIMLFLFV |
| UACG | CXWCSRSGRPATSFLFIMIVLHQHDEDNKNYXY |
| UAGC | STPAGRSRSRWCXCFLFLMILVDEDEQHQHNKNKXYXY |
| UGAC | SAPTNKNKQHQHDEDEXYXYFLFLVLIMISRSRGXCWC |
| GUAC | ASPTKNKQHQXYXEDEGXCWRSRMILFLV |
| AUCG | RSRSCWCXGRPASTKNKNYXYXEDEDHQHQLVLFLFIMI |
| AUGC | TSPAGRXCWRSRKNKXYXQHQEDEVLFLMI |
| AGUC | TAPSLFLVIMIKNDEQHYXWCRGSR |
| **GAUC** | ATPSLFLMIVGSRCXWXYHQKNDE |
| CUAG | RXWCGRSNKEDYXQHPSATIMIVFL |
| **CUGA** | HQXYNKEDGRSCXWRPSTAVMIFL |
| CGUA | HQHDEDNKNYXYFLFIMIVLPATSCWXCSRSGR |
| GCUA | EDEDHQHQKNKNYXYXLFLFIMLVGRSRSCXCWSTPA |
| CAUG | RSRSGCWXCFLFVIMILPTASYXYDEDNKNHQH |
| CAGU | LMIFLFLVGWCXCSRSRPTSADEDEXYXYNKNKQHQH |
| CGAU | LVFLFLIMINKNKXYXYDEDEQHQHPASTSRSRXCWCGR |
| GCAU | VLFLIMKNKXYXQHQEDEGRWCXRSRTSPA |
| ACUG | RSRSRGCWCXLFLFVLIMIKNKNHQHQEDEDYXYXSAPT |
| ACGU | IMLFLVGWCXRSRKNKQHQXYXEDEASPT |
| AGCU | IMIVLFLPSATKNDEXYHQRCWXGSR |
| GACU | VIMLFLPSTAGSRWXCRQHYXKNDE |

Two sets of data are included in Table 5.2, one in which the first codon position changes most slowly and the third fastest (Table 5.2a) and one in which the second position is assigned to the slowest changing index (Table 5.2b). In each of these, adjacent identical residues have been reduced to a single entry.

# 5.5 MULTIPLE SEQUENCE ALIGNMENTS

The discussion in Section 5.3 described how general substitution preferences can be extracted from the consideration of pairwise sequence alignments. These matrices can then be reapplied to sequences to gain some insight into the mutational preference of specific residues in specific proteins. With multiply aligned sequences, however, much more can be learned about a specific position if the generalized matrix is avoided and the raw data itself is analysed. This requires, for the particular protein of interest, having several homologous sequences aligned which must be sufficiently diverse (but can still be correctly aligned). The various residues seen aligned at a position then represent a snap-shot or sample of the possible variation allowed at that point.

## 5.5.1    General principles

The most apparent property of a position is how well it has been conserved rel-ative to other positions. High conservation is a good indicator that the position is vital to the protein, being required for either a structural or functional (or sometimes a folding) purpose. Since the hydrophobic residues are involved in packing in the core, their degree of conservation correlates with their degree of burial – an observation that can be used in the construction of model tertiary structures (Aszódi & Taylor, 1994a,b). In the less conserved hydrophobic positions, the appearance of acids with long side-chains is indicative of a posi-tion more on the edge of the core where the longer side-chains can extend their terminal polar atoms out into the solvent. A typical pattern of this type would be the substitution of an otherwise conserved Phe with a Tyr. Benner (1989) has compiled a practical guide to the application of many of these 'rules' which can be applied 'by hand' to a multiple sequence alignment, with a view to extracting structural information.

High conservation of hydrophilic amino acids is usually indicative of a func-tional role for the conserved acids, either binding or catalytic, but, as pointed out by Benner (1989), care must be exercised as substrate specificity can vary between subfamilies that have been aligned together.

Multiple sequence alignments can help resolve the ambiguous nature of Gly. If an otherwise conserved Gly position is seen to show some shift towards the hydrophilics (Asp, or Ser) or Pro then a surface role would be anticipated, while if some shift is seen towards the hydrophobics, burial would be more probable.

Variation over a wide variety of hydrophilic residues (including Gly and Pro), and especially insertions and deletions in the alignment, are indicative of a surface location (where there are few steric constraints). Often, the only amino acids missing from such sites are the large hydrophobic residues. In membrane proteins, variation is also indicative of exposure to solvent, but for these proteins the solvent is hydrophobic and so are the variable positions (Taylor et al., 1994).

## 5.5.2    Colouring multiple alignments

When assessing multiple sequence alignments it is useful to 'highlight' the con-served features. While this is often done using whatever colour of fluorescent marker pens that are available, it is useful to have a more systematic approach.

Many colours might be (and have been) used for the amino acids, but there is one reasonably unique assignment (Taylor, 1997b) shown in Table 5.3. The order of the acids in Table 5.3 gives little scope for change that does not separate some closely linked pair or triple (ST, LIV, FYW, RK, NQ, DE). The greatest ambiguity is found in the orange acids, where any permutation of the four acids would be acceptable. That chosen keeps the oxygen-containing Ser

**Table 5.3** The colours of the amino acids. Each acid (left) is assigned a systematic colour as: Xish-Xey-X, where X represents at most two colours combined in proportions, Xish < Xey < X. A trivial name is also given (in parentheses) along with the RGB (red, green, blue) weights. (N.B., magenta and cyan, which are normally the complements of green and red, have been slightly displaced from their conventional assignments). The order is circular, as indicated by repeated entries above and below the lines.

| Amino acid | Three-letter code | | Colour | | RGB |
|---|---|---|---|---|---|
| Glutamine | Gln | RBR | Redish-bluey-red | (magenta) | 1.0 0.0 0.8 |
| Glutamic acid | Glu | BRR | Blueish-redey-red | (violet) | 1.0 0.0 0.4 |
| Aspartic acid | Asp | RRR | Redish-redey-red | (red) | 1.0 0.0 0.0 |
| Serine | Ser | YRR | Yellowish-redey-red | (scarlet) | 1.0 0.2 0.0 |
| Threonine | Thr | RYR | Redish-yellowey-red | (vermillion) | 1.0 0.4 0.0 |
| Glycine | Gly | YRY | Yellowish-redey-yellow | (orange) | 1.0 0.6 0.0 |
| Proline | Pro | RYY | Redish-yellowey-yellow | (tangerine) | 1.0 0.8 0.0 |
| Cysteine | Cys | YYY | Yellowish-yellowey-yellow | (yellow) | 1.0 1.0 0.0 |
| Alanine | Ala | GYY | Greenish-yellowey-yellow | (lemon) | 0.8 1.0 0.0 |
| Valine | Val | YGY | Yellowish-greeney-yellow | (lemon-lime) | 0.6 1.0 0.0 |
| Isoleucine | Ile | GYG | Greenish-yellowey-green | (lime) | 0.4 1.0 0.0 |
| Leucine | Leu | YGG | Yellowish-greeney-green | (grass) | 0.2 1.0 0.0 |
| Methionine | Met | GGG | Greenish-greeney-green | (green) | 0.0 1.0 0.0 |
| Phenylalanine | Phe | BGG | Blueish-greeney-green | (emerald) | 0.0 1.0 0.4 |
| Tyrosine | Tyr | GBG | Greenish-bluey-green | (turquoise) | 0.0 1.0 0.8 |
| Tryptophan | Trp | BGB | Blueish-greeney-blue | (cyan) | 0.0 0.8 1.0 |
| Histidine | His | GBB | Greenish-bluey-blue | (peacock) | 0.0 0.4 1.0 |
| Arginine | Arg | BBB | Blueish-bluey-blue | (blue) | 0.0 0.0 1.0 |
| Lysine | Lys | RBB | Redish-bluey-blue | (indigo) | 0.4 0.0 1.0 |
| Asparagine | Asn | BRB | Blueish-redey-blue | (purple) | 0.8 0.0 1.0 |
| Glutamine | Gln | RBR | Redish-bluey-red | (magenta) | 1.0 0.0 0.8 |
| Glutamic acid | Glu | BRR | Blueish-redey-red | (violet) | 1.0 0.0 0.4 |
| Aspartic acid | Asp | RRR | redish-redey-red | (red) | 1.0 0.0 0.0 |
| Serine | Ser | YRR | yellowish-redey-red | (scarlet) | 1.0 0.2 0.0 |
| Threonine | Thr | RYR | redish-yellowey-red | (vermillion) | 1.0 0.4 0.0 |

and Thr close to the red acidic residues. The ordering also has some useful groupings; for example:

- Hydrophobic amino acids are green (GYY–BGB).
- Aromatic amino acids are greeney-blue (BGG–GBB).
- Amino acids found in loops are red and orange (RRR–RYY).
- Large polar acids are purple and blue (GBB–RBR).

The colours of the acids are all fully saturated and can be used to colour a multiple sequence alignment simply by averaging the RGB (red, green, blue) components of each aligned acid. Such averaging results in lower saturation, giving more delicate pastel shades for mixed positions and eventually white for positions at which all acids are equally represented (or for which there is a balanced selection). Simultaneously, the degree of conservation can be emphasized by using intensity to darken less conserved positions giving more browns and muddy colours where there is little conservation and eventually black where nothing is conserved. Gaps in the alignment fit well into this scheme and can be treated as black amino acids. (See Taylor, 1997b, for other interesting possibilities.)

## 5.5.3   Coordinated mutations

It has often been thought that a spontaneous mutation in a protein might be compensated by a change at another site that lies adjacent in the structure. For example, the appearance of a negative charge might be balanced by a positive charge appearing nearby, or the loss of a Cys might leave a disulphide partner free to mutate. Many workers have searched for this effect (Altschuh *et al.*, 1987a; Taylor & Hatrick, 1994; Neher, 1994) and despite some optimistic results (Gobel *et al.*, 1994), the effect has never been demonstrated to the extent where it would prove to be of use in providing constraints on possible tertiary structure. This is unfortunate since, if such results could be obtained, they would provide three-dimensional constraints directly from sequence data.

There may be a glimmer of hope in looking at specific protein–protein interactions (Altschuh *et al.*, 1987b; Vernet *et al.*, 1992; Pazos *et al.*, 1997) but within a single globular domain, a recent simulation suggests that the background noise from chance associations is too high to allow any useful constraints to be extracted (Pollock and Taylor, 1997; Pollock *et al.*, 1999). The reason for this can be found in the number of interactions made by a single amino acid. If one position changes, then there are many neighbours able to spread and absorb the change, without any particular one coming under strong pressure to change. This contrasts with the situation in RNA where the interactions are much more one-to-one, and in this molecule correlated changes provide a useful method to predict structure (Stormo, 1990).

## 5.5.4   Extrapolating variation

For the general principles outlined above to become apparent, it is necessary to have a large number of sequences aligned (covering a wide range of variation). While such data are becoming increasingly available (with the genomes of a broad phylogenetic range of organisms currently being determined), there is still often not enough sequences available to be useful and it is necessary to try and extrapolate to the full range of variation from the observed sample. Different solutions to this problem have been attempted, ranging from the very simple to very complex statistical treatments. Complexity really begins when 'proper' probabilities of acid occurrence are required and when the degree of relatedness among the sample sequences is also considered. These additional difficulties will not be considered here.

### 5.5.4.1   Amino acid sets

One of the simplest ways to predict 'missing' amino acids that might be possible substitutions at a site, is to categorize the physicochemical properties of those that have been observed and extend the match-set to include all others with similar properties. An early attempt to do this used a Venn diagram (Fig. 5.5).

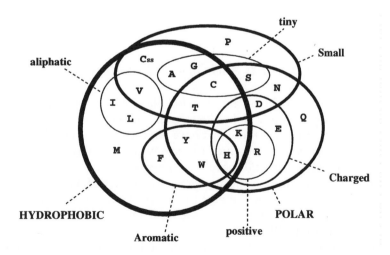

**Figure 5.5**   Venn diagram grouping properties of the amino acids. The acids are grouped into property sets, of which the largest are Hydrophobic and Polar (Pro is excluded from these because of its unique stereochemistry). Within these major groupings, the subsets of size, charge and aromaticity are distinguished. Cys is represented twice with Css indicating the disulphide linked from cystene.

Having noted the combination of sets that contained the observed acids, then all others also found in the same combination of property sets constitute possible substitutions (Taylor, 1986). While useful, this approach had the theoretical problem that the definition of set boundaries is somewhat arbitrary and also the practical problem that a misalignment of the sequences (or a sequence error) could quickly extend any set to encompass all 20 amino acids.

### 5.5.4.2   Substitution matrix

The above problems can be avoided by taking an amino acid exchange matrix as the model of what acids should be allowed. If a position, say contained a Val and an Ala, the score (or propensity, since the values are not probabilities) for another acid (X) to appear would be $M_{AX} + M_{VX}$ (where $M_{XY}$ is the substitution value for acids X and Y, given some Dayhoff-like matrix $M$). Clearly, this approach will work well given just one observed acid (which is what the matrix was constructed from). However, if there are a lot of sequences then the substitution 'space' (or distribution) for that position may already be well sampled and the use of a matrix will only blur the useful constraints that have been observed.

Some control can be retained over excessive 'spreading' of the distribution by using iteration combined with renormalization of the scores (Taylor, 1997a). By resetting the original mean of the scores after each iteration the positive scores are cut-back each time and cannot spread over all acids (Fig. 5.6).

### 5.5.4.3   Pseudocounts and Dirichlet mixtures

Given a frequency count of amino acids observed at a position, a related approach is simply to add 'fake' (or pseudo) amino acid counts to the frequency vector to represent those that have not been seen or may be under-represented. This approach can adjust to the level of sampling and add a lot of fake counts if only a few acids have been observed or add only a few counts if there are many sequences in the alignment. The degree of addition of pseudocounts to the vector can be estimated by some clever statistical analysis that can also incorporate the degree of relatedness among the sequences (Altschul et al., 1989; Altschul, 1991).

Where pseudocounts are based upon the sum of individual estimates from each amino acid observed, it is possible to use a more sophisticated statistical technique (a Dirichlet mixture) that effectively simultaneously considers all the amino acids observed by incorporating information from their joint distribution. One can imagine a situation where this might contribute: for example, if either Val and Ile are seen at a position, then their individual distributions might both admit Leu as a pseudocount. However, a position with both Val and Ile might be in a β-sheet and therefore be less likely to admit Leu (which favours α-helix). Although such structural information is not explicitly specified in Dirichlet mixes, some equivalent will be implicit in the joint distribution.

```
GDAEAAAKTS   *..**.*...*..7..2.**.....    ADEGKST ---> ADEGKNQST
EEDEEEDDDQ   ...**.........1..*........    DEQ ---> DENQ
WWFKKRKKKA   *....*....*.......*0...*..    AFKRW ---> AFKRSW
QNDSSATSAA   *..*4.0......*..*.**.....    ADNQST ---> ADEGNQST
QHAATANVRL   *......*0..*0*..**1*.*...    AHLNQRTV ---> AHILMNQRSTV
VVVVIIVVVV   ..........*..........*...    IV ---> IV
LLLTSKKKKK   ..........**1.....**.....    KLST ---> KLMST
NGKAAAAAAS   *..2..*...*..*....*2.....    AGKNS ---> ADGKNST
VICLVLAFLS   *.*..*..*..*4.....*3.*...    ACFILSV ---> ACFILMSTV
WWWWWWWWWW   ....................*..     W ---> W
GAGGGGGGDE   *..**.*......2....1......    ADEG ---> ADEGNS
KKPKKKKKKE   ....*.....*....*2........    EKP ---> EKPQ
VVVVVIVIIF   .....*..*..20.........*...    FIV ---> FILMV
EEENNDGSEN   2..**.*......*....*.....    DEGNS ---> ADEGNS
APAVIVAGGA   *.....*.*......*..22.*...    AGIPV ---> AGIPSTV
```

```
GDAEAAAKTS   *..**.*...*..8.14.**.....    ADEGKST ---> ADEGKNPQST
EEDEEEDDDQ   ...**..2.....3..*........    DEQ ---> DEHNQ
WWFKKRKKKA   *....*....*......*4...*..    AFKRW ---> AFKRSW
QNDSSATSAA   *..*7.21..1..*.0*.**.....    ADNQST ---> ADEGHKNPQST
QHAATANVRL   *..21..*3.1*3*.2**4*.*...    AHLNQRTV ---> ADEHIKLMNPQRSTV
VVVVIIVVVV   ........*..11........*...    IV ---> ILMV
LLLTSKKKKK   0........**52....**.....    KLST ---> AKLMNST
NGKAAAAAAS   *..52.*...*..*....*5.....    AGKNS ---> ADEGKNST
VICLVLAFLS   *.*..*..*..*8.....*6.*...    ACFILSV ---> ACFILMSTV
WWWWWWWWWW   ................2....*..     W ---> RW
GAGGGGGGDE   *..**.*......5..2.4......    ADEG ---> ADEGNQS
KKPKKKKKKE   1..1*.....*..3.*5.1......    EKP ---> ADEKNPQS
VVVVVIVIIF   ......*..*..53.........*..0    FIV ---> FILMVY
EEENNDGSEN   4..**.*......*..1.*1.....    DEGNS ---> ADEGNQST
APAVIVAGGA   *.....*.*......*..44.*...    AGIPV ---> AGIPSTV
```

**Figure 5.6** Amino acid match-set extension. A sequence alignment of 10 globins (over a short region near the amino terminus) runs vertically on the left. This is recoded as a matrix showing the weight on each amino acid in alphabetical order (A–Y). Amino acids present at the aligned position are marked by an asterisk and have weight 1, while a weight less than zero is represented by a dot. Intermediate weights are indicated as the integral part of 10 times their value. The upper panel has little set extension and the lower panel has greater set extension.

A detailed analysis of Dirichlet mixtures, pseudocounts and other measures has shown that the Dirichlet mixture is the best at reconstituting the 'true' (underlying) distribution as defined by a maximum-entropy measure (Karplus, 1995).

## REFERENCES

Altschuh D., Lesk A.M., Bloomer A.C. & Klug A. (1987a) *Prot. Eng.* **1**, 228.
Altschuh D., Lesk A.M., Bloomer A.C. & Klug A. (1987b) *J. Molec. Biol.* **193**, 693–707.
Altschul S.F. (1991) *J. Molec. Biol.* **219**, 555–565.
Altschul S.F., Carroll R.J. & Lipman D.J. (1989) *J. Molec. Biol.* **207**, 647–653.
Aszódi A. & Taylor W.R. (1994a) *Biopolymers* **34**, 489–506.
Aszódi A. & Taylor W.R. (1994b) *Prot. Eng.* **7**, 633–644.
Baker E.N. & Hubbard R.E. (1984) *Prog. Biophys. Mol. Biol.* **44**, 97–179.
Benner S.A. (1989) *Adv. Enzyme Reg.* **28**, 219–236.
Bowie J.U., Lüthy R. & Eisenberg D. (1991) *Science* **253**, 164–170.
Crick F.H.C. (1968) *J. Molec. Biol.* **38**, 367–379.
Dayhoff M.O., Schwartz R.M. & Orcutt B.C. (1978) In *Atlas of Protein Sequence and Structure*, (Dayhoff, M.O., ed.) Volume 5, Suppl. 3. Natl. Biomed. Res. Foundation, Washington DC, pp. 345–352.
Fischer D., Wolfson H., Lin S.L. & Nussinov R. (1994) *Prot. Sci.* **3**, 769–778.
French S. & Robson B. (1983) *J. Molec. Evol.* **19**, 171–178.
Gibson T.J. & Lamond A.I. (1990) *J. Molec. Evol.* **30**, 7–15.
Gobel U., Sander C., Schneider R. & Valencia A. (1994) *Prot. Struct. Funct. Genet.* **18**, 309–317.
Gonnet G.H., Cohen M.A. & Benner S.A. (1992) *Science* **256**, 1443–1445.
Henikoff S. & Henikoff J.G. (1992) *Proc. Natl. Acad. Sci. USA* **89**, 10915–10919.
Jones D.T., Taylor W.R. & Thornton J.M. (1992) *Comp. App. Bio. Sci.* **8**, 275–282.
Jones D.T., Taylor W.R. & Thornton J.M. (1994) *FEBS Lett.* 269–275.
Karplus K. (1995) In *The Third International Conference on Intelligent Systems for Molecular Biology (ISMB)*, Rawlings C., Clark D., Altman R., Hunter L., Lengauer T. & Wodak S., (eds) AAAI Press, Menlo Park, pp. 188–196.
Kyte J. (1995) *Structure in Protein Chemistry*. Garland Publishing, New York and London.
Lamond A.I. & Gibson T.J. (1990) *Trends Genet.* **6**, 145–149.
Martin B.L. (1997) In *Encyclopedia of Molecular Biology and Molecular Medicine*, Vol. 5, Meyers, R.A., (ed.), VCH, New York, pp. 49–56.
McLachlan A.D. (1971) *J. Molec. Biol.* **61**, 409–424.
Mehta P.K., Heringa J. & Argos P. (1995) *Prot. Sci.* **4**, 2517–2525.
Neher E. (1994) *Proc. Natl. Acad. Sci. USA* **91**, 98–102.
Overington J., Šali A., Johnson M.S. & Blundell T.L. (1990) *Proc. Roy. Soc. Lond. B* **241**, 132–145.
Overington J., Donnelly D., Johnson M.S., Šali A. & Blundell T.L. (1992) *Prot. Sci.* **1**, 216–226.

Pazos F., Helmer-Citterich M., Ausiello G. & Valencia A. (1997) *J. Molec. Biol.* **271**, 511–523.

Pollock D.D., Taylor W.R. & Goldman N. (1999) *J. Molec. Biol.*, **287**, 187–198.

Pollock D.D. & Taylor W.R. (1997) *Prot. Eng.* **10**, 647–657.

Rice D.W. & Eisenberg D. (1997) *J. Molec. Biol.* **267**, 1026–1038.

Sneath P.H.A. (1966) *J. Theor. Biol.* **12**, 157–195.

Stormo G.D. (1990) In *Molecular Evolution; Computer Analysis of Protein and Nucleic Acid Sequences*, Doolittle, R.F. (ed.), Vol. 183. *Meth. Enzymol.* Academic Press, San Diego, pp. 211–221.

Swanson R. (1984) *Bull. Math. Biol.* **46**, 187–204.

Swindells M.B., MacArthur M.W. & Thornton J.M. (1995) *Nature Struct. Biol.* **2**, 596–603.

Taylor W.R. (1986) *J. Theor. Biol.* **119**, 205–218.

Taylor W.R. (1989) *Prog. Biophys. Molec. Biol.* **54**, 159–252.

Taylor W.R. (1997a) *J. Molec. Biol.* **269**, 902–943.

Taylor W.R. (1997b) *Prot. Eng.* **10**, 743–746.

Taylor W.R. & Hatrick K. (1994) *Prot. Eng.*, **7**, 341–348.

Taylor W.R. & Jones D.T. (1993) *J. Theor. Biol.* **164**, 65–83.

Taylor W.R., Jones D.T. & Green N.M. (1994) *Prot. Struct. Funct. Genet.* **18**, 281–294.

Thorne J.L., Goldman N. & Jones D.T. (1996) *Molec. Biol. Evol.* **13**, 666–673.

Vernet T., Tessier D., Khouri H.E. & Altschuh D. (1992) *J. Molec. Biol.* **224**, 461–471.

Vingron M. & von Haeseler A. (1997) *J. Comp. Biol.* **4**, 23–34.

Yamashita M.M., Wesson L., Eisenman G. & Eisenberg D. (1990) *Proc. Natl. Acad. Sci. USA* **87**, 5648–5652.

# 6     Sequence comparison

## Michael Gribskov

San Diego Supercomputer Center, San Diego, CA, USA

## 6.1   INTRODUCTION

Sequence comparison is one of the most important activities in computational molecular biology. Once a pair of sequences has been determined to be homologous, sequence comparisons allow a precise map to be created, associating specific residues in one sequence with the other. This permits one to map functional information such as the location of secondary structural elements, domains, active sites, and regulatory regions from a well-studied molecule to a new sequence. This has been one of the most powerful processes in molecular biology, allowing years or often decades of biochemical investigation to be immediately applied to a newly sequenced molecule. Initially, this mapping of functional information was performed on a one sequence at a time basis. The availability of many entire genomes resulted in sequence comparison methods becoming the basis of functional annotation of genomic sequences (see Ouzounis *et al.*, 1996, for example). Because this automatic functional annotation is the only annotation available for many genes, it is important to understand both the basic methods, and how to confirm the annotation for oneself.

    Construction of an informational map requires that the molecules under investigation be homologous, that is that they share a common ancestor. Let us briefly examine why this is so. Consider the active site of an enzyme. An active site is constructed of several amino acid residues with the required chemical properties. The residues must be placed and oriented rather precisely in three-dimensional space to perform their catalytic function, but there are no constraints on where the residues might lie in the polypeptide chain. There are a virtually unlimited number of ways the polypeptide chain might be arranged, with varying spacing between the critical residues, in order to achieve the required three-dimensional positioning required for activity. One can see, then, that there is no reason for unrelated protein molecules to have similar sequences, even if they share precisely the same enzymatic mechanism. Only the fact that the molecules share a relatively recent common ancestor estab-

*Genetics Databases*
ISBN 0-12-101625-0

lishes a relationship between the sequences and the structures and functions of the molecules. It is this relationship that allows us to conclude that molecules are homologous when we see statistically significant sequence similarity. It is this same relationship, and the fact that one can discern that the molecules once had identical structure and function that allows us to create the informational map that we are seeking.

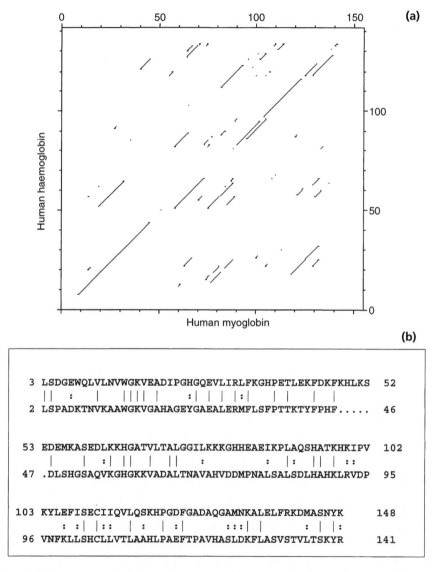

**Figure 6.1**   Dot matrix (a) and alignment (b) representations of the matching between human α-haemoglobin and human myoglobin.

The initial determination that two molecules are homologous is generally made based on the results of a database search (see Chapter 1). The statistical evaluation of database searches is critical because the inference of homology is critically dependent on this result. Statistics are somewhat less important in sequence comparisons because one typically already knows that the molecules are homologous. The task is to construct the greatest match between the molecules in order to produce an informational map.

There are two basic types of matching that can be used as the basis of informational map: dot matrix plots or dotplots, and alignments (Fig. 6.1). Each representation produces equivalent information and has its own advantages.

## 6.2  DOTPLOTS

Dot matrix plots, or dotplots, have been in use in molecular biology for 30 years (Fitch, 1966, 1969; Gibbs and McIntyre, 1970; McLachlan, 1971; Maizel & Lenk, 1981). Dotplots remain a useful and powerful means of comparing sequences. The power of dotplots lies in their use of an extremely sensitive pattern detection tool, the human eye, to evaluate graphically the similarity between sequences. Dotplots have a number of advantages over sequence alignments: they are insensitive to rearrangements and perform better in the presence of low entropy (simple) or duplicated sequences. Also, since they have no penalties for gaps, they can easily detect similar regions separated by long insertions.

As homologous sequences diverge, they gradually accumulate point mutations and, more slowly, insertions and deletions. Homologous sequences will have many stretches where identical bases or similar residues can be matched up without gaps when the two sequences are juxtaposed. This produces a striking pattern of diagonal lines in a dotplot, as can be seen in Fig. 6.1.

The goal of a dotplot is to construct a graphical matrix. The elements in each row correspond to one base or residue in the first sequence and the elements in each column correspond to a base or residue in the second. At any location where the row element and column element match, a dot is placed in the graph. Regions of the sequences that can be matched without introducing gaps generate a contiguous series of dots along a diagonal line; these contiguous dots are often joined together into a line.

When a dotplot is constructed in this simple way, the plot is very noisy. This is because many of the single base or residue matches between the two sequences are random. The matches that represent homology tend to occur consecutively in the two sequences and hence would lie along diagonal lines in the dot matrix plot. This pattern of diagonal matches representing homology can be strengthened by calculating the number of matches within a window.

For example, one might use a window of seven bases and place a dot in the graph only when four of the seven bases are identical. In this case, the dot would typically be placed in the position corresponding to the centre of the window.

While it is sufficient with nucleic acid sequences to simply score the number of identical bases, proteins require a more complicated system. The 20 amino acid residues have many overlapping chemical similarities. For example, in one situation an alanine residue may be considered to be similar to a serine residue owing to its similar size. In another situation, an alanine may be similar to a leucine owing to its nonpolar side-chain (Chapter 5). These complicated chemical similarities require the use of a 20×20 scoring table that gives a score for each possible pairwise comparison. Many scoring tables are available, but the most commonly used tables derive from empirically counting the numbers of amino acid substitutions within homologous families of proteins. The PAM (per cent accepted mutation) matrices of Dayhoff *et al.* (1978) have been widely used as scoring tables for both dotplots and alignments. More recently, BLOSUM (blocks' substitution) matrices introduced by Henikoff & Henikoff (1992, 1993) have gained favour. Both of these scoring systems provide a range of scoring tables appropriate for different evolutionary distances. For the PAM series, these are typically referred to by names such as PAM-40, PAM-120, or PAM-250, which represent 40 mutations, 120 mutations, and 250 mutations respectively per 100 residues. Although PAM-250 was originally recommended, a PAM-120 matrix is probably better for general use (Altschul, 1991). BLOSUM matrices also span an evolutionary range: BLOSUM-90 represents a relatively short divergence, and BLOSUM-20 would represent relatively long divergence. A BLOSUM-62 matrix is most often used in sequence comparisons (Henikoff & Henikoff, 1993).

## 6.2.1    Insertion and deletion

Insertions and deletions between sequences appear as vertical or horizontal offsets in the diagonal lines indicating blocks of similar residues. The size of the offset is the length of the insertion/deletion. Dot matrix methods are able to detect all sizes of gaps, even very long ones, because they do not specifically penalize gaps.

## 6.2.2    Duplications

Duplications appear as sets of parallel diagonal lines. When a sequence is compared with itself, these diagonal lines will fill in a square region of the plot (Fig. 6.2a). When related sequences are compared, the region of parallel diagonal

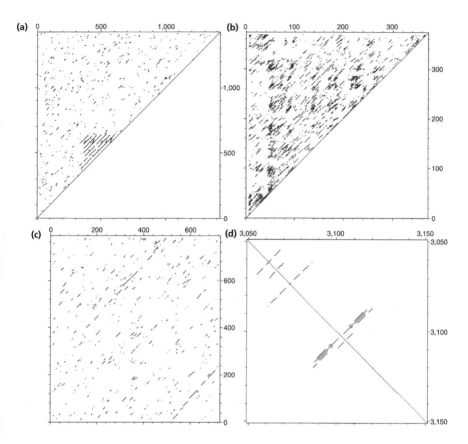

**Figure 6.2** Examples of dot matrix plots. (a) Self comparison of the human death-associated protein kinase sequence; note the eight 28 residue ank repeats around residues 373–637. (b) Self comparison of the human δ-opioid receptor. Transmembrane segments are located at residues 85–102, 125–144, 175–150, 216–238, 262–284, and 294–310. (c) Comparison of yeast *p*-aminobenzoate synthetase (vertical) and rhizobium anthranilate synthetase. These evolutionarily related enzymes show a rearrangment of the glutamine amidotransferase domain, residues 1–228 and 515–718 in yeast and rhizobium respectively, and the *p*-aminobenzoate synthetase/anthranilate phosphoribosyl transferase domain, residues 304–788 and 44–445 respectively. (d) Inverted repeats at the end of the *Escherichia coli* RNA polymerase α operon. The inverted repeats at bases 3063–3084 and 3088–3117 presumably function as termination signals.

lines will fill a rectangle if the number of repeated units is different in the two sequences. Alignment programs typically perform quite poorly in the presence of unequal numbers of repeating units, so a dotplot is an important method for both detecting and analysing this situation. Both the number of repeating units and the length of the repeat may be found from a dotplot. The number of repeats is the number of diagonal lines and the length of the repeat is the spacing between the parallel lines.

Low-entropy sequences, for instance those composed of monomer or dimer repeats, are common in biological sequences. Some of these sequences, such as poly-A sequences in mRNAs have known biological functions. Others appear to arise as a byproduct of a variety of genetic events such as insertion and excision of transposons and gene conversion. These sequences can be thought of as a special case of a repeated sequence wherein the repeat is very short, one or two bases or residues. Such sequences will appear as a dense block of dots in a self-comparison, and may cause a repeating pattern of 'blobs' in both self-comparisons and binary comparisons, especially when a low threshold for displaying matching regions is used. A commonly seen example is in the comparison of transmembrane proteins in which the highly hydrophobic transmembrane regions produce a plaid-like pattern in low-stringency dotplots (Fig 6.2b).

## 6.2.3    Rearrangements

Rearrangements appear in dotplots as inversions of diagonal matching regions along one axis of the plot (Fig. 6.2c). Rearrangements are fairly common in nucleic acids, although detecting them may be difficult owing to the low sensitivity of nucleic acid sequence comparisons. Except for domain shuffling and certain cyclic permutations, rearrangements are relatively rare in protein sequences. Nevertheless, it is wise to check for such features with a dotplot because dynamic programming sequence alignments (see below) are completely unable to detect them.

## 6.2.4    RNA structure

RNA secondary structures, commonly called stems, are formed from inverse-complementary sequences. Such sequences may be easily seen in dotplots by using a scoring system that scores for the appropriate complementary pairs, AU, GC, and possibly GU. Alternatively, a dotplot of the sequence versus the inverse-complementary sequence may be made (see Quigley *et al.*, 1984, for some examples). In either case, the potential stem regions appear as diagonal regions at right angles to a two-fold symmetry axis (Fig. 6.2d). Because such plots cannot accurately include energetic considerations such as the destabilizing effect of loops, or neighbour dependence of base stacking energies, they can give only a qualitative picture of possible structures. However, they do give a global picture of possible structures, not only those that would be part of a predicted minimum free energy structure. This can be an advantage in certain situations.

## 6.2.5   Practical hints

The major controllable parameters in dotplots are usually the window size and the threshold for producing a dot. For proteins, the amino acid residue scoring matrix can often also be specified. For protein sequences, especially distantly related sequences, the window cannot be made too long. This is because the presence of insertions and deletions limits the ability to find the diagonal regions. A window of 10–20 residues is a good place to start. For nucleic acids, one typically uses a longer window because the four-letter nucleic acid alphabet allows many more diagonal regions to be found at random. Windows of 30 or more bases are common, and windows in the hundreds may be useful when large sequences are compared. The signal to noise ratio in a dotplot is an important consideration. As the sequences being compared grow longer, the number of windows being compared grows as the square of the length of the sequences, i.e., quadratically. However, the windows that represent homology only grow linearly. This implies that one should use an increasingly stringent threshold as the sequences get longer. A good practical rule is to make plots that have three to five times as many dots as the length of the sequences (e.g., 3000–5000 dots for a 1000 base sequence).

## 6.3   ALIGNMENTS

Sequence alignments are a one-to-one matching of the bases or residues in two sequences. Alignments conventionally are shown with each sequence on a separate line, padded as necessary with 'blank' characters to indicate gaps, and often with an intervening line indicating the quality of the match (Fig 6.1). Computer-generated sequence alignments typically show the optimal, or highest probability, alignment between the two sequences. Dynamic programming sequence alignments (Needleman & Wunsch, 1970; Smith & Waterman, 1981) are mathematically guaranteed to report one of the optimal alignments considering all numbers, positions, and sizes of gaps. While it is not critical to know the details of the alignment algorithm, it is important to understand the mathematical definition of optimal because it may not precisely agree with what a biologist might consider optimal.

The alignment problem boils down to a question of how many gaps need to be inserted in each sequence, where they should be placed, and how long they must be. To determine the best answer to this problem, the computer requires a scoring function that specifies the quality of a match. This scoring function takes into account scores for both the matched residues or bases and gaps:

$$\text{Alignment quality} = \Sigma\text{Score}_{\text{match}} + \Sigma\text{Score}_{\text{gap}}.$$

The score for matched residues (bases), $\text{Score}_{\text{match}}$, is typically a similarity measure. It is positive for residues that are similar and negative for residues that are different with respect to chemistry or evolution. The score for gaps, $\text{Score}_{\text{gap}}$, is therefore negative; the addition of gaps always decreases the quality of an alignment. The gap score is usually defined by the affine gap cost model. Simply, this means the (negative) score for a gap is a linear function of the length of the gap, and that the function has a non-zero intercept:

$$\text{Score}_{\text{gap}} = G_{\text{init}} + G_{\text{extend}} \times \text{length}.$$

$G_{\text{init}}$ can be thought of as the cost of gap initiation or the cost of inserting the first blank character across from a base or residue in one sequence. $G_{\text{extend}}$ can be thought of as the cost of gap extension or the cost of inserting each additional blank character in an already initiated gap. The reasons for using this function for gap penalties are computational: it makes the alignment calculation much faster and it has been found that an affine gap cost is necessary to find correct alignments (Fitch & Smith, 1983; Vingron & Waterman, 1994; Gusfield & Stelling, 1996). However, even brief consideration would suggest that such a model for gap costs is inherently nonbiological. The affine gap cost model implies that insertions/deletions are exponentially less likely as their length increases. This makes sense if long insertions and deletions occur as a series of single base insertions. However, long insertions and deletions are quite likely to occur as single events, and the affine gap model excessively penalizes these events.

Almost all alignment programs allow the user to define the gap initiation and gap extension parameters. Because the gap penalties determine the trade-off between allowing a 'bad' match, i.e. one with a negative score, and inserting a gap, these parameters critically influence the resulting alignment. If the gap penalties are too low, very large numbers of gaps are inserted and almost any sequences can be aligned with a high proportion of identical bases or residues. If the gap penalties are too high, no gaps can be inserted, and the alignment is equivalent to simply sliding the sequences past each other until a maximum score is achieved. Since there is currently no way, a priori, to determine the best values for the gap penalties, it is essential to try a variety of different values and examine the effect on the resulting alignment.

## 6.3.1   Variations on the basic approach

There are a number of slightly different algorithms and implementations of the basic dynamic programming alignment approach. The most significant differences lie in the difference between global and local approaches. The original alignment algorithms (Needleman & Wunsch, 1970) were global. That is, every base or residue in both sequences must be either matched with another base or residue, or matched to a blank (thereby incurring a gap penalty). Global algorithms are still widely used, especially in phylogenetic estimation. They have been replaced by local alignment (Smith & Waterman, 1981) algorithms for most applications where the existence of a match needs to be established.

Local alignment algorithms find the best matching segment between two sequences, not necessarily using all bases or residues of either sequence. Local alignment algorithms are widely preferred because they are capable of finding and matching domains or motifs, even in the presence of large amounts of unrelated sequence. Local alignment approaches generally have a higher signal to noise ratio than global alignments because they focus on the most closely related regions and omit regions that are not appreciably different from random. Some common implementations of alignment programs provide the option of not penalizing gaps that occur at the beginning or end of either sequence (non end-weighted). This has an effect somewhat similar to a local alignment because large insertions and deletions can be made at the ends without cost. It is better to use a local alignment algorithm that trims off the unconserved end regions than use a non end-weighted method that will align the unconserved end region with an internal sequence.

## 6.3.2   Significance of alignments

One frequently wants to ask whether a given alignment is significant. In practical terms, this amounts to asking whether the alignment score is significantly higher than one would expect for random sequences. Because there are no adequate theoretical treatments that describe the distribution of scores for alignments with gaps, significance is determined empirically.

The traditional method of determining alignment significance is to use a Monte Carlo approach. One or both sequences are randomized while preserving its overall base or residue composition. The alignment is repeated a number of times, typically hundreds or thousands, empirically generating a set of scores typical of random sequences of the same length and composition. The mean and standard deviation of this distribution are then used to convert the original alignment score to a Z–score or standardized score:

$$Z = (\text{Score}_{\text{alignment}} - \text{Average Score}_{\text{random}}) / \text{Standard Deviation}_{\text{random}}.$$

The Z-score should be relatively independent of the scoring system, gap penalties, alignment length and other parameters allowing one to choose the parameter settings to produce the most significant alignment. Traditionally, Z > 6.0 is taken as supporting an inference of homology, and Z< 3.0 as suggesting a lack of homology. Intermediate values are ambiguous.

It is believed that the score distribution for alignments with gaps is similar to that for maximal segment pairs (best aligned regions without gaps). A theoretical description of the score distribution for maximal segment pairs was given by Karlin & Altschul (1990). This distribution is a type of extreme value or Gumbel distribution. It is possible to fit data for alignments of either random sequences or unrelated sequences (e.g., from a database search) to such a theoretical distribution and to determine the significance of an alignment (Pearson & Lipman, 1988). A straightforward graphical procedure was described by Collins & Coulson (1990). In this approach, the alignment scores for random sequences are plotted against the log of the fraction of alignments with equal or higher scores (Fig 6.3). This plot generally shows a linear relationship and allows one to calculate the P-value, the probability of observing a random alignment score as high or higher as the one obtained with the sequences of interest, by simply looking up the score on the regression line. The P-value of the alignment is a much better way of determining the significance of an alignment than a simple Z-score.

In addition to asking whether the overall alignment is significant, one may also ask which parts of the matching are most likely to be correct. This question cannot be answered in a simple way because most software only displays

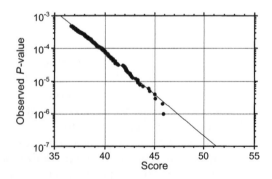

**Figure 6.3**   Calculation of P-value for dynamic programming alignment. The alignment score for randomized sequences with the same scoring matrix and gap penalties is plotted versus the observed fraction of alignments with equal or higher score (observed P-value).

a single most significant alignment (there may be others with this score) and none of the sub-optimal alignments. The alignment produced is also highly dependent on the scoring matrix and gap penalties. One can investigate this question in a qualitative way by using a variety of gap penalties and observing which portions of the alignment are stable with respect to changes in the gap penalties, and which portions are sensitive to small changes in parameters. The stable regions are the most confident regions, while the unstable regions may represent only one of several nearly equivalent ways to match that region. This investigation can be greatly helped by comparison with a dotplot in which the various regions can be matched to conserved blocks of sequence.

### 6.3.3  Relationship to database searches

Database searches are often thought of as approximations to alignments. This is unfortunate, because the alignment problem is quite different from a database search. In a database search, one is interested in finding sequences that are significantly more similar than unrelated sequences. This is because one is trying to support an inference of homology, which can then be used to map functional information from one sequence to another. Because there are many more unrelated sequences than related sequences in the database, a very stringent threshold (e.g., $P<10^{-5}$) must be used to ensure that unrelated sequences are not accepted as homologous. This often limits the alignment to only a short very similar segment (consider the BLAST program, for instance). In a typical alignment, one has already decided that the sequences are homologous. Therefore one is willing to reduce the significance thresholds in order to produce the alignment that uses the largest amount of the sequences under investigation.

## 6.4  MOTIF-BASED APPROACHES

In addition to pairwise alignments, sequences are often compared with descriptions of motifs or domains that derive from known groups of sequences. These approaches produce the same kind of informational map produced by a traditional pairwise comparison.

### 6.4.1  Regular expressions

Regular expressions describe a motif in terms of the bases or residues observed in known sequences. These kinds of descriptions are generally very good at detecting the motif in previously known sequences, but much poorer in detecting slightly different versions of the motif in newly discovered sequences. This is because the motif description has no provision for generalization (for matching to characters not previously observed), nor is there any adequate way to statistically evaluate a partial match. One knows that a match to, for example, five out of six positions, is much more common than a match to six out of six, but it is difficult to say just how much more common. This, in turn, makes it difficult to decide just how much faith to place in such a partial match.

The most common and useful regular expression motif descriptions are those found in the PROSITE (Bairoch *et al.*, 1997) dictionary of sequence 'signatures'. Signatures are regular expression type descriptions, usually about 20 residues long, often centred on an area of functional significance. An example of a signature and an explanation of the syntax is shown in Fig 6.4. PROSITE provides a wealth of information in addition to the signature, including a brief description of the known or inferred biological function of the proteins

```
ID    ASN_GLN_ASE_1; PATTERN.
AC    PS00144;
DT    APR-1990 (CREATED); NOV-1997 (DATA UPDATE); NOV-1997 (INFO UPDATE).
DE    Asparaginase / glutaminase active site signature 1.
PA    [LIVM]-x(2)-T-G-G-T-[IV]-[AGS].
NR    /RELEASE=35,69113;
NR    /TOTAL=21(21); /POSITIVE=13(13); /UNKNOWN=0(0); /FALSE_POS=8(8);
NR    /FALSE_NEG=0; /PARTIAL=0;
CC    /TAXO-RANGE=A?EP?; /MAX-REPEAT=1;
CC    /SITE=6,active_site;
DR    P18840, ASG1_ECOLI, T; P38986, ASG1_YEAST, T; P00805, ASG2_ECOLI, T;
DR    P43843, ASG2_HAEIN, T; P11163, ASG2_YEAST, T; P30363, ASPG_BACLI, T;
DR    P26900, ASPG_BACSU, T; P06608, ASPG_ERWCH, T; Q60331, ASPG_METJA, T;
DR    Q10759, ASPG_MYCTU, T; P50286, ASPG_WOLSU, T; P10172, ASPQ_ACIGL, T;
DR    P10182, ASPQ_PSES7, T;
DR    Q04945, ADHB_CLOAB, F; Q50637, APT_MYCTU , F; Q65914, FIBP_ADECT, F;
DR    P30941, IAP2_SOLTU, F; P17979, IAP_SOLTU , F; Q03197, ICTC_SOLTU, F;
DR    P16348, ICTD_SOLTU, F; P55908, YCGA_BACSU, F;
3D    3ECA; 4ECA; 1WSA; 1AGX; 3PGA; 4PGA;
DO    PDOC00132;
```

**Figure 6.4**   PROSITE signature for the asparaginase–glutaminase active site. The signature itself is given in the PA line using the following syntax: each position in the signature is separated by a hyphen; single letters specify required matching residues; square brackets enclose a group of allowed residues, parentheses, give a repeat count, e.g. x(2) means any two residues. (Curly brackets, not used above, indicate residues that are not allowed.)

The DR lines list the SWISS-PROT accession numbers and entry names of the known positive (T) and false positive (F) sequences that match this signature. Sequences may also be listed as N, a false negative which belongs to the family but does not match the signature, P, a partial sequence that probably belongs to the family but is missing the region of the signature, and ?, a sequence whose status is unknown.

containing the motif. There are listings of all known proteins thought to have the same function and whether or not they match the motif. The extensive annotation in PROSITE makes it a powerful tool for functional annotation of novel sequences. Furthermore, PROSITE plays an important role in providing classified data that can be used to test other methods.

## 6.4.2 Motifs and profiles

A more general method for describing sequence motifs relies on position-specific scoring matrices (PSSMs, Fig 6.5). A PSSM may be thought of as corresponding to a multiple sequence alignment. Each position on the PSSM corresponds to a column in the multiple alignment and the values represent the bases or residues allowed at that position. These values are often represented as a log-odds score comparing the logarithm of the likelihood of observing the base or residue at random, divided by the likelihood of observing it at that position by chance. The score for comparing a sequence and a PSSM is simply the sum of the scores for the bases or residues in the sequence at each position in the PSSM. The use of PSSMs in sequence comparison is common, and often not obvious. For example, programs such as Psi-BLAST (Altschul *et al.*, 1997) and ClustalW (Higgins *et al.*, 1996) both use PSSMs.

Explicit score distributions have been derived for matching without gaps to PSSMs (Bailey & Gribskov, 1997). These relations allow the accurate calculation of *P*-values, and therefore evaluation of the significance for BLOCKS (Henikoff *et al.*, 1998) and MEME (Bailey and Elkan, 1994). Matching with gaps to PSSMs suffers from the same statistical difficulties as aligning sequences with gaps: there is no adequate theory predicting the distribution of scores. As with sequence alignments, matching to such motif descriptions can be evaluated by a Monte Carlo procedure, or by empirical fitting to a Gumbel-type distribution.

| Cons | A | C | D | E | F | G | H | I | K | L | M | N | P | O | R | S | T | V | W | Y | Init | Ext |
|---|---|---|---|---|---|---|---|---|---|---|---|---|---|---|---|---|---|---|---|---|---|---|
| I | -17 | -53 | -53 | -43 | -9 | -50 | -55 | 76 | -49 | 37 | 26 | -42 | -41 | -41 | -44 | -39 | -12 | 52 | -107 | -34 | 100 | 100 |
| L | -25 | -67 | -57 | -46 | 5 | -53 | -44 | 33 | -48 | 70 | 39 | -45 | -37 | -29 | -44 | -43 | -24 | 24 | -88 | -20 | 100 | 100 |
| A | 53 | -48 | -9 | -10 | -63 | 18 | -44 | -33 | -35 | -54 | -36 | -7 | 0 | -28 | -44 | 18 | 20 | -14 | -82 | -61 | 100 | 100 |
| T | 17 | -48 | -24 | -31 | -67 | -22 | -50 | -17 | -22 | -55 | -25 | -2 | -15 | -39 | -45 | 27 | 77 | -16 | -86 | -62 | 100 | 100 |
| G | 2 | -92 | -15 | -27 | -77 | 83 | -77 | -78 | -59 | -83 | -82 | -14 | -38 | -54 | -83 | -1 | -32 | -48 | -117 | -101 | 100 | 100 |
| G | 2 | -92 | -15 | -27 | -77 | 83 | -77 | -78 | -59 | -83 | -82 | -14 | -38 | -54 | -83 | -1 | -32 | -48 | -117 | -101 | 100 | 100 |
| T | 17 | -48 | -24 | -31 | -67 | -22 | -50 | -17 | -22 | -55 | -25 | -2 | -15 | -39 | -45 | 27 | 77 | -16 | -86 | -62 | 100 | 100 |
| I | -17 | -49 | -49 | -40 | -10 | -49 | -55 | 85 | -45 | 21 | 21 | -39 | -42 | -43 | -41 | -38 | -9 | 53 | -100 | -33 | 100 | 100 |

**Figure 6.5** A position-specific scoring matrix (PSSM). This is a log-odds matrix corresponding to the asparaginase–glutaminase active site shown in Fig. 6.4. This PSSM is a profile; the two columns on the far right give weights (in per cent) to be applied to gap initiation and extension penalties.

### 6.4.3    Hidden Markov models

Hidden Markov models (HMMs) are a more general model for describing sequence patterns than a PSSM (Baldi *et al.*, 1994; Krogh *et al.*, 1994). HMMs explicitly describe both the match states, corresponding to the aligned columns in the multiple sequence alignment, and the expected distribution of residues in the padding or insert states. HMMs also include explicit probabilities for deletions at each point in the multiple sequence alignment. HMMs allow more sophisticated models than the simple linear model embodied in a PSSM. However, these features are relatively rarely used for proteins and there is a strong similarity between HMMs and profiles in this case. In practice, one must typically have a large number of sequences, hopefully hundreds, with at least an approximate alignment before beginning the HMM training. The matching of a trained HMM to a sequence is performed similarly to the matching of two sequences. However, the total probability of all possible matches is often evaluated rather than the probability of the optimal match. Most of the comments on evaluating dynamic programming alignments (apart from the effect of gap penalties that are built into the HMM) apply. Because of the position-specific gaps, matching to HMMs must be evaluated empirically, much like PSSMs or profiles.

### 6.4.4    Motif databases

Many databases are now available for matching to motif descriptions. These include PROSITE (Bairoch *et al.*, 1997), BLOCKS (Henikoff & Henikoff, 1996) and Prints (Attwood *et al.*, 1998). As we approach the time when virtually all protein sequences are known, these resources will become increasingly complete, and hence increasingly able to attach useful functional annotation to newly determined sequences.

## 6.5    CONCLUSION

Sequence comparison can be an extremely powerful technique if properly performed. Many hours of laboratory work may be saved by a well-performed sequence comparison but many hours of computer and laboratory time can be wasted if the job is poorly done. As a final suggestion here is a protocol for matching:

- Start with a query and template sequence that are clearly homologous. This requires a database or alignment $P$-value of $10^{-5}$ or lower, or compelling biological evidence.

- Compare the query sequence with itself using a dot matrix analysis. Identify low-entropy sequences and repeated sequences. Consider breaking the sequence up into segments if these features are seen.
- Compare the query and template sequences using a dot matrix analysis. Again, pay special attention to repeated sequences and potential rearrangements.
- Align the query and template sequences using a dynamic programming alignment procedure. The sequences may have to be broken up into parts if there are repeated sequences or rearrangements. Examine many different gap opening and extension penalties and examine the effect on the alignment. Compare the most stable regions with the dotplots.
- Compare the query sequence to motif/domain databases. In favourable cases this may establish a function for the conserved regions seen in dotplots and alignments. Multiple sequence alignment can also be very useful.

## REFERENCES

Altschul S.F. (1991) *J. Mol. Biol.* **219**, 555–565.
Altschul S.F., Madden T.L., Schäffer A.A., Zhang J., Zhang Z., Miller W. & Lipman D.J. (1997) *Nucl. Acids Res.* **25**, 3389–3402.
Attwood T.K., Beck M.E., Flower D.R., Scordis P. & Selley J. (1998) *Nucl. Acids Res.* **26**, 304–308.
Bailey T.L., & Elkan C. (1994). *Proc. 2nd Int. Conf. Intelligent Sys. Molec. Biol.*, pp. 28–36.
Bailey T.L. & Gribskov M. (1997) *J. Comput. Biol.*, **4**, 45–59.
Bairoch A., Bucher P. and Hofmann K. (1997) *Nucl. Acids Res.* **25**, 217–221.
Baldi P., Chauvin Y., Hunkapiller T. & McClure M.A. (1994) *Proc. Natl. Acad. Sci. USA* **91**, 1059–1063.
Dayhoff M.O., Schwartz R.M. & Orcutt B.C. (1978) In *Atlas of Protein Sequence and Structure*, Dayhoff, M.O. (ed.). National Biomedical Research Foundation, Washington DC., pp. 345–352.
Collins J. & Coulson A. (1990). *Meth. Enzymol.* **183**, 474–487.
Fitch W.M. (1966) *J. Mol. Biol.* **16**, 9–16.
Fitch, W.M. (1969) *Biochem. Genet.* **3**, 99–108.
Fitch W.M. & Smith T.F. (1983) *Proc. Natl. Acad. Sci. USA* **80**, 1382–1286.
Gibbs A.J. & McIntyre G.A. (1970) *Eur. J. Biochem.* **16**, 1–11.
Gusfield D. & Stelling P. (1996) *Meth. Enzymol.* **266**, 481–494.
Henikoff S. & Henikoff J.G. (1992) *Proc. Natl. Acad. Sci. USA* **89**, 10915–10919.
Henikoff S. & Henikoff J.G. (1993) *Proteins* **17**, 49–61.
Henikoff J.G. & Henikoff S. (1996) *Meth. Enzymol.* **266**, 88–105.
Henikoff S., Pietrokovski S. & Henikoff J.G. (1998) *Nucl. Acids Res.* **26**, 309–312.

Higgins D.G., Thompson J.D. & Gibson T.J. (1996). *Meth. Enzymol.* **266**, 383–402.

Karlin S. & Altschul S.F. (1990). *Proc. Natl. Acad. Sci. USA* **87**, 2264–2268.

Krogh A., Brown M., Mian I.S., Sjölander K. & Haussler D. (1994). *J. Mol. Biol.* **235**, 1501–1531.

Maizel J.V. & Lenk R.P. (1981) *Proc. Natl. Acad. Sci. USA* **78**, 7665–7669.

McLachlan A.D. (1971) *J. Mol. Biol.* **61**, 409–424.

Needleman S.B. & Wunsch C.D. (1970) *J. Mol. Biol.* **48**, 443–453.

Ouzounis C., Casara G., Sander C., Tamames J., & Valencia A. (1996) *TIBTECH 14*, 280–285.

Pearson W.R. & Lipman D.J. (1988) *Proc. Natl. Acad. Sci. USA* **85**, 2444–2448.

Quigley G.J., Gehrke L., Roth D.A. & Auron P.E. (1984) *Nucl. Acids Res.* **121**, 347–366.

Smith T.F. & Waterman M.S. (1981) *J. Mol. Biol.* **147**, 195–197.

Vingron M. & Waterman M.S. (1994) *J. Mol. Biol.* **235**, 1–12.

# 7 Simple Repetitive Sequences in DNA Databanks

*Jörg T. Epplen & Eduardo J.M. Santos*

Molecular Human Genetics, Faculty of Medicine, Ruhr University, Bochum, Germany

## 7.1 INTRODUCTION

A characteristic feature of all eukaryotic and most prokaryotic genomes is the presence of various amounts of different repetitive DNA sequences (Epplen *et al.*, 1998). By definition, repetitive elements occur more than once per haploid genome. A subclass, tandemly organized repetitive DNA sequences, represents a major component of all eukaryotic genomes. A broadly accepted classification scheme of tandemly repetitive sequences is derived from the lengths of their repetition units (see Tautz, 1993). (1) The longest category, classical satellites, may comprise several thousand base pairs. Satellites were named in the early space-exploration era (for a review see Britten and Kohne, 1968). Since classical satellites were identified in buoyant density gradients before the DNA sequencing methodology grew efficient, this repetitive sequence category has, in the meantime, been defined as consisting of quite heterogeneous sequence entities. The more perfect the tandem repeats of given lengths are organized in a satellite, the less amenable is this DNA block to complete sequence analysis. The degree of sequence redundancy amounts to up to $10^7$ at each satellite locus. There are only one or a few loci per genome of a given eukaryotic species that contribute to the satellite(s). Satellites are mostly situated in heterochromatic regions, e.g. at the centromeres or on the sex chromosomes. (2) The basic units of minisatellites (Jeffreys *et al.*, 1985a, b) are nine to approximately 100 base pairs long and two to several hundred copies exist per locus. For example, in the human genome, many thousands of minisatellites are interspersed with practically all other DNA elements but they are often clustered in sub-telomeric chromosomal regions (Royle *et al.*, 1988). (3) Simple repeat motifs range from one

*Genetics Databases*
ISBN 0-12-101625-0

to six bases in length with five to approximately 100 perfect tandem copies at each locus. Simple repeats can be scattered throughout the genomes in, e.g. $10^5$ copies in the human genome, a number that corresponds fairly well to the gene number in this species. Since the early 1990s, simple repeats have often been called microsatellites (as the next smaller repeat motif entity to minisatellites; Weber & May, 1989; Litt & Luty, 1989), after the enzymatic amplification by polymerase chain reaction (PCR) became exploitable. After the advent of PCR, it quickly became clear that simple sequence loci have a great potential for many aspects of genome research (Tautz, 1989): DNA profiling, genome mapping and, more recently, population genetics. The large amount of research on these sequences has inevitably led to a large number of different concepts and theories about the nature of simple repeats, but none of them has clarified the truly significant parts of the phenomenon. In order to maintain a certain degree of homogeneity in this chapter and because of space limitations cryptically simple sequences (Tautz *et al.*, 1986) and tandemly repeated genes, as well as additional (partly ill-defined) classes of tandem repeats, are not covered here.

Besides their proven tool character, is there any biological function for tandem repetitive DNA sequences? Pure simple repeats have earlier been suggested to represent the building blocks of a primordial protein-coding sequence in the primitive code early during the evolution of life (Ohno & Epplen, 1983). In addition, apparently the entire evolution of eukaryotes has been accompanied by such repetitive DNA. One or the other simple repetitive tandem tract appears to be present in every genome but the more complex and longer repetitive elements are not found in all of the different species. So far, the true biological meaning of this type of genomic redundancy is far from being perceived and understood. Probably therefore, the original 'junk hypothesis' (Ohno, 1972) has now widely been accepted, neglecting the many initial reservations and the disinterest for it. Here, we concentrate on the application aspect of microsatellites for population genetics. Many applications have already been reported and far-reaching conclusions have been drawn, for example on hotly debated issues such as the confirmatory microsatellite evidence for early population expansion in Africa (Shriver *et al.*, 1997) or the dispersion hypothesis of human Y chromosome haplotypes in global populations, which is based exclusively on five microsatellites (Deka *et al.*, 1996). This chapter is written for the non-expert user of computers including those who have to send email enquiries for their DNA database searches (EBI FASTA searches, http://srs.ebi.ac.uk/; NCBI BLAST searches, http://www.ncbi.nlm.nih.gov/; EMBL BLAST 2 searches, http://www.bork.embl-heidelberg.de/).

## 7.2   MICROSATELLITES IN DATABASES FOR POPULATION GENETIC ANALYSES

An ever-increasing number of microsatellite loci has been identified inadvertently during sequencing specific genes in genomic DNA or isolated and developed specifically for various purposes over the recent years; see compilations in the World Wide Web (WWW) servers at Généthon (http://www.genethon.fr/), GDB (http://gdb.infobiogen.fr/) or CHLC (http://www.chlc.org/). In addition, several large-scale genome sequencing efforts contribute increasingly to the databank representation of simple repeat sequences. For the individual FASTA searches of the complete EMBL/GenBank databank (>$1.5 \times 10^9$ base pairs in >$2.1 \times 10^6$ sequence entries as of February 21, 1998), all perfect simple repeat stretches 40 bases long were recorded (see Table 7.1). At present, the frequencies of different simple repeats in the complete EMBL/GenBank may be markedly skewed. $(GT)_n/(AC)_n$ repeats especially are outstandingly abundant. These elements have for a long time been the backbone of many genome mapping efforts, yet most $(GT)_n/(AC)_n$ microsatellites from the human genome map are not entered, as such, in the EMBL/GenBank databases, thus reducing their preponderance. On the other hand, mapping has shifted more and more to the employment of tri- and tetra-nucleotide repeats (and such procedures will include nonrepetitive biallelic polymorphisms) for technical reasons of easier separation of the neighbouring alleles. Furthermore, pure poly(A)$^+$ blocks are currently the second most abundant motifs often representing byproducts of sequencing cDNAs to mRNAs or frequent retroposon events involving reverse transcribed pseudogenes in the genomes. $(AT)_n/(TA)_n$, $(ATA)_n/(TAT)_n$ and $(AATA)_n/(TATT)_n$ repeats may have been derived frequently from original poly(A)$^+$ tails, especially when incorporated into other repetitive elements such as the most abundant short interspersed nucleotide elements (SINES), i.e. Alu repeats in human (see e.g. Yandava *et al.*, 1997). The predominance of $(GATA)_n/(TATC)_n$ elements cannot be explained convincingly or independently without considering the apparent advantages of perfect dinucleotide periodicities (see below). Simple repeat-containing sequences from humans make up no more than 10–20% of all of the databank entries from all other organisms.

The recent short-term development of the simple repeat representation in the EMBL/GenBank data bank also can be followed in Table 7.1. The poly(A)$^+$ blocks especially increased in frequency significantly faster recently than all the other simple repeats, thus stressing the contributions of the gene expression projects to the DNA data banks (Boeke, 1997). Therefore, the presence of the different motifs depends also on the research strategies employed that finally lead to the submission of the given sequence information. Although these frequencies may be somewhat biased (especially since sequences from Man outnumber those of all other species in the combined databanks), a few rules for the overall evolutionary success of simple repeats can be formulated on the basis of their existence in present-day genomes:

(1) The sequence composition is decisive for the representation of long simple repeats; W (A/T) >> S (G/C) (long A/T rich simple repeats far outnumber G/C rich elements)

**Table 7.1**    Representation of simple repetitive DNA sequences in the
EMBL/GenBank data bank (40 bases and longer).

| Simple repeat motif | EMHUM | | GEALL | |
|---|---|---|---|---|
| | 1.1.1998 | 28.4.1996 | 21.11.1997 | 21.2.1998 |
| $(G)_{40}$ | 0 | 7 | 120 | 139 |
| $(A)_{40}$ | **187** | **917** | **4109** | **5247** |
| $(G)_{20}(A)_{20}$ | 0 | 0 | 0 | 6 |
| $(G)_{20}(T)_{20}$ | 0 | 2 | 2 | 2 |
| $(G)_{20}(C)_{20}$ | 0 | 0 | 0 | 0 |
| $(A)_{20}(G)_{20}$ | 0 | 2 | 3 | 10 |
| $(A)_{20}(T)_{20}$ | 0 | 0 | 0 | 0 |
| $(T)_{20}(A)_{20}$ | 0 | 0 | 0 | 0 |
| $(T)_{20}(G)_{20}$ | 0 | 0 | 0 | 1 |
| $(C)_{20}(G)_{20}$ | 0 | 0 | 0 | 9 |
| $(GA)_{20}$ | 95 | **624** | **1270** | **1447** |
| $(GT)_{20}$ | **1379** | **3994** | **6064** | **6654** |
| $(GC)_{20}$ | 0 | 0 | 0 | 0 |
| $(AT)_{20}$ | **206** | **526** | **1382** | **1816** |
| $(GT)_{10}(GA)_{10}$ | 82 | 239 | 403 | 435 |
| $(GA)_{10}(GT)_{10}$ | 43 | 62 | 93 | 96 |
| $(AG)_{10}(GT)_{10}$ | 0 | 1 | 1 | 1 |
| $(GA)_{10}(TG)_{10}$ | 0 | 0 | 0 | 0 |
| $(GT)_{10}(AG)_{10}$ | 0 | 2 | 3 | 3 |
| $(TG)_{10}(GA)_{10}$ | 0 | 1 | 3 | 4 |
| $(GT)_{10}(AT)_{10}$ | 32 | 59 | 107 | 122 |
| $(AT)_{10}(GT)_{10}$ | 12 | 43 | 82 | 93 |
| $(GT)_{10}(TA)_{10}$ | 0 | 0 | 0 | 0 |
| $(TG)_{10}(AT)_{10}$ | 0 | 2 | 2 | 2 |
| $(TA)_{10}(GT)_{10}$ | 2 | 3 | 8 | 8 |
| $(AT)_{10}(TG)_{10}$ | 0 | 0 | 0 | 0 |
| $(GA)_{10}(AT)_{10}$ | 0 | 0 | 0 | 0 |
| $(AT)_{10}(GA)_{10}$ | 0 | 1 | 1 | 1 |
| $(AT)_{10}(AG)_{10}$ | 29 | 11 | ? | 100 |
| $(GA)_{10}(TA)_{10}$ | 0 | 1 | 1 | 1 |
| $(AG)_{10}(TA)_{10}$ | 0 | 0 | 0 | 0 |
| $(TA)_{10}(AG)_{10}$ | 0 | 0 | 0 | 0 |
| $(GC)_{10}(AT)_{10}$ | 0 | 0 | 0 | 0 |
| $(GC)_{10}(TA)_{10}$ | 0 | 0 | 0 | 0 |
| $(AT)_{10}(CG)_{10}$ | 0 | 0 | 0 | 0 |
| $(TA)_{10}(CG)_{10}$ | 0 | 0 | 0 | 0 |
| $(GGA)_{13.3}$ | 5 | 30 | 52 | 54 |
| $(GGT)_{13.3}$ | 5 | 19 | 30 | 32 |
| $(GGC)_{13.3}$ | 12 | 16 | 28 | 29 |
| $(GAA)_{13.3}$ | 12 | 75 | 164 | 191 |

| Simple repeat motif | EMHUM | | GEALL | |
| --- | --- | --- | --- | --- |
| | 1.1.1998 | 28.4.1996 | 21.11.1997 | 21.2.1998 |
| $(GAT)_{13.3}$ | 5 | 26 | 38 | 48 |
| $(GAC)_{13.3}$ | 0 | 3 | 6 | 6 |
| $(GTA)_{13.3}$ | 1 | 23 | 31 | 34 |
| $(GTT)_{13.3}$ | 6 | 41 | 80 | 91 |
| $(GCA)_{13.3}$ | 39 | 86 | 187 | 222 |
| $(AAT)_{13.3}$ | 44 | **238** | **418** | **485** |
| $(GGGA)_{10}$ | 7 | 20 | 41 | 46 |
| $(GGGT)_{10}$ | 1 | 1 | 2 | 2 |
| $(GGGC)_{10}$ | 0 | 0 | 0 | 0 |
| $(GGAA)_{10}$ | 102 | **272** | **494** | **555** |
| $(GGAT)_{10}$ | 34 | 76 | 183 | 189 |
| $(GGTA)_{10}$ | 0 | 1 | 7 | 12 |
| $(GGTT)_{10}$ | 1 | 20 | 18 | 17 |
| $(GGCA)_{10}$ | 1 | 10 | 19 | 21 |
| $(GAGT)_{10}$ | 1 | 2 | 5 | 8 |
| $(GAAA)_{10}$ | 136 | **399** | **822** | **957** |
| $(GAAT)_{10}$ | 5 | 20 | 23 | 25 |
| $(GAAC)_{10}$ | 0 | 0 | 1 | 1 |
| $(GATA)_{10}$ | 109 | **1685** | **2061** | **2119** |
| $(GATT)_{10}$ | 0 | 17 | 15 | 15 |
| $(GACA)_{10}$ | 1 | 28 | 41 | 50 |
| $(GACT)_{10}$ | 0 | 5 | 1 | 1 |
| $(GTAA)_{10}$ | 0 | 4 | 8 | 8 |
| $(GTAT)_{10}$ | 17 | 65 | 110 | 134 |
| $(GTTA)_{10}$ | 1 | 0 | 1 | 1 |
| $(GTTT)_{10}$ | 2 | 10 | 24 | 26 |
| $(GCAA)_{10}$ | 0 | 7 | 11 | 20 |
| $(GCAT)_{10}$ | 0 | 0 | 20 | 20 |
| $(GCTA)_{10}$ | 0 | 0 | 0 | 0 |
| $(GCCA)_{10}$ | 0 | 2 | 2 | 2 |
| $(AAAT)_{10}$ | 115 | **131** | **400** | **479** |
| $(AATT)_{10}$ | 0 | 6 | 4 | 4 |
| $(GA)_{10}(GACA)_5$ | 2 | 4 | 12 | 11 |
| $(GA)_{10}(GATA)_5$ | 0 | 2 | 2 | 3 |
| $(GT)_{10}(GACA)_5$ | 0 | 0 | 1 | 1 |
| $(GT)_{10}(GATA)_5$ | 0 | 0 | 0 | 0 |
| $(GACA)_5(GA)_{10}$ | 0 | 2 | 2 | 2 |
| $(GACA)_5(GT)_{10}$ | 0 | 0 | 0 | 0 |
| $(GATA)_5(GA)_{10}$ | 0 | 1 | 2 | 2 |
| $(GATA)_5(GT)_{10}$ | 0 | 0 | 0 | 0 |
| $(GATA)_5(AAT)_{6.7}$ | 0 | 0 | 0 | 0 |
| $(AAT)_{6.7}(GATA)_5$ | 0 | 0 | 0 | 0 |
| $(GAAA)_5(GATA)_5$ | 0 | 0 | 1 | 1 |
| $(GATA)_5(GAAA)_5$ | 1 | 1 | 1 | 1 |

(2) Perfect dinucleotide periodicity warrants evolutionary success for simple repeats.

(3) Long CpG dinucleotide containing simple repeat motifs are exceedingly rare.

It is known from many vertebrates, that CpG dinucleotides can be methylated (Jabbari *et al*., 1997) and that they represent hotspots for mutations (Cooper & Krawczak, 1989), thus efficiently eliminating longer perfect $(CG)_n/(CG)_n$ tracts. Interestingly, the second rule outlined above applies also to combined sequence motifs consisting of two adjacent blocks of different motifs, e.g. $(GT)_n/(AC)_n \cdot (GA)_m/(TC)_m$ tracts. Whenever the perfect dinucleotide periodicity is interrupted – even only at the border of the two perfect tracts – the combined blocks exceedingly rarely grow to the size tested here (see $(GT)_n/(AC)_n \cdot (AG)_m/(CT)_m$ or $(TG)_n/(CA)_n \cdot (GA)_m/(TC)_m$ and all combined dinucleotide blocks without completely continuous periodicity). Simple trinucleotide repeats such as $(CAG)_n/(CTG)_n$ have been searched for particularly intensely in many medically oriented projects. These trinucleotide blocks are certainly over represented in Table 7.1. The other trinucleotide motifs will be dealt with separately below (see Section 5). In summary, the sequence content of the truly abundant simple repeat motifs is highly reminiscent of the AT/GC content of spacer and intron DNA in many eukaryotes. The mere abundance of most of these repeated genomic elements cannot now be explained satisfactorily without implicating elusive imminent genome mechanisms. These facts render the simple repeats puzzling objects for future investigations.

Several features of these loci such as Mendelian inheritance, high mutation rates and easy screening by PCR make them powerful tools for population genetic studies (in particular for approaches within a given species). The critical point to reveal a realistic evolutionary picture of a species is to obtain appropriate genetic similarity measures. First, an adequate number of loci has to be employed to avoid high variance values on the measure (Zhivotovsky & Feldman, 1995). Therefore, the necessity of databases of microsatellites for population studies is evident. Some databases are kept at the Smithsonian Laboratory of Molecular Systematics (http://nmnhgoph.si.edu/gophermenus/MicrosatelliteDatabase.html) and at the Human Population Laboratory at Stanford University (http://lotka.stanford.edu). Second, the genetic similarity measure used must be appropriate, considering the mutation model (stepwise or infinite allele model) as well as the high mutation rates of the microsatellites. Recently, a great number of genetic distance measures have been proposed that take these factors into consideration. Other demographic aspects of the populations (such as gene flow and effective population size) must be considered to obtain accurate estimates of genetic distances since these factors determine the speed of genetic differentiation between populations. In the next section, procedures for estimating genetic distances, their applicability to microsatellite data bases and their reliability with respect to the aforementioned demographic factors are discussed. A list of Web sites is included allowing one to obtain software to compute several of the statistics mentioned in this section.

## 7.3 GENETIC DISTANCES

The high mutation rates of microsatellites (Weber & Wong, 1993) require to take the molecular details of the mutation processes into consideration. The polymerase slippage mechanism has been accepted to explain the frequent stepwise mutations of these loci (Levinson & Gutman, 1987; Schloetterer & Tautz, 1992). According to this model, a given allele with 10 repeat units does not mutate directly to an allele length with, for example, 16 units. Instead, intermediate states with 11, 12, 13, 14 and 15 repeats have to be passed through. This model of mutation is called the stepwise mutation model (SMM). However, mutations involving several repeat units have been inferred from multimodal distributions of allele frequencies. Modal peaks sometimes constitute allelic classes and no intermediate size alleles are observed between these classes, thus invalidating the stepwise mutation assumption (Epplen *et al.*, 1997; Santos *et al.*, 1998). Other molecular features of the microsatellite mutational process, such as asymmetry of mutation distribution and repeat range constraint, remain equally probable (Amos *et al.*, 1996; Primmer *et al.*, 1996). In this scenario, a number of genetic distance estimates have been developed. A group of estimates considers the difference in repeat lengths between alleles to take information about the genetic differentiation under the SMM. Two related estimates may be cited: $R_{ST}$ and $\rho_{ST}$ (Slatkin, 1995; Rousset, 1996).

As an example let us consider three hypothetical populations X, Y and Z, each with two alleles of a given microsatellite locus. Table 7.2 shows the repeat length of the alleles encountered as well as their respective frequencies. If we consider only allele frequencies the genetic distances between populations X and Y will be equal to the distance between X and Z. However, if we consider the allelic relationships under assumption of the SMM, X becomes genetically more similar to Y than to Z.

**Table 7.2** Allele lengths and their frequencies in three hypothetical populations.

| Repeat length | Population X | Population Y | Population Z |
|---|---|---|---|
| 12 | 0.8 | 0.7 | 0.7 |
| 13 | 0.2 | | |
| 14 | | 0.3 | |
| 15 | | | 0.3 |

$R_{ST}$ (Slatkin, 1995) may be defined as $R_{ST} = (S - S_W)/S$. In practice, $S_W$ is twice the average of the variances in repeat length within each population and S is twice the variance in repeat length in the amalgamation of all populations. Unbiased estimators should be used to obtain the estimated variance. Thus, the greater the differentiation between the populations investigated, the higher will be the value of S, consequently increasing $R_{ST}$.

The analysis of molecular variance (Michalakis & Excoffier, 1996) may be applied to microsatellite loci and an estimator $\rho_{ST}$ may be obtained (Rousset, 1996). Basically, $\rho_{ST}$ is an analogue of the $F_{ST}$, as defined by Weir & Cockerham (1984), considering evolutionary relationships among the alleles. The distance between alleles is entered as a matrix of pairwise Euclidean distances (Michalakis & Excoffier, 1996). Hence if the distances between all different alleles is equal to one, $\rho_{ST}$ will equal $F_{ST}$. But if the distance between alleles varies, $\rho_{ST} \neq F_{ST}$. There is also a close relationship between $\rho_{ST}$ and $R_{ST}$:

$$\rho_{ST} = (1-c)R_{ST}/1-cR_{ST}$$

where $c = (2N-P)/(2NP-P)$.

P and N are the number of populations sampled and the total number of individuals sampled, respectively (Rousset, 1996). Related genetic distance measures were reported and more information about theses latter distance measures are contained in the reports of Goldstein *et al.* (1995) and Goldstein & Pollock (1997).

There are close relationships between all aforementioned genetic distance measures. Thus, users of any genetic distance estimate that consider a strict stepwise mutation model must pay attention to a number of points. Obviously, one must observe if the loci investigated follow a SMM. Most empirical data (Weber & Wong, 1993, Amos *et al.*, 1996; Santos *et al.*, 1998) and theoretical studies (Valdes *et al.*, 1993; Di Rienzo *et al.*, 1994) suggest a general SMM for microsatellites. However, several microsatellites have multimodal allele frequency distributions, sometimes separated by several repeat units. Each peak of the allele frequencies may constitute a class (Epplen *et al.*, 1997; Santos *et al.*, 1998). Such allele distributions may be explained by earlier rare mutation events involving several repeat units (Freimer & Slatkin, 1997; Santos *et al.*, 1998). Alternatively, gene flow may introduce alleles into a population and the locus does not obligatorily follow the SMM. In intraspecific analyses multimodal allele frequency distributions and the presence of allelic classes are strong arguments against the use of these genetic distance measures. In this case, other measures should be preferred that do not necessitate to fulfil the assumptions of the SMM.

Another important group of genetic distances are centred around the sum of the products of shared alleles between populations. Thus, this group does not assume a particular model of evolution (Goldstein & Pollock, 1997). The distances Das, Da and Dc may be described by the following general formula:

$$D = c \{1-[Sum(XiYi)]^a\}^b.$$

Xi and Yi are the frequencies of allele i in populations X and Y, respectively; $a$, $b$ and $c$ are equal to one for D$as$, 0.5, 1 and 1 for D$a$ and 0.5, 0.5 and 2Sqrt(2)/π for Dc (Goldstein & Pollock, 1997). For multiple loci the average distance over all loci is used. These distances are most accurate for recently separated populations (Takezaki & Nei, 1996). However, the degree of allele sharing may be influenced by sampling effects, in particular for small sample sizes (Goldstein & Pollock, 1997).

Recently, several software programs have been developed to estimate genetic distances for application of microsatellite data. These may be accessed via the Internet (Table 7.3).

**Table 7.3** Address, name and description of the software to estimate genetic distances and other relevant statistics on population genetics using microsatellite loci.

| Address | Programs | Description |
| --- | --- | --- |
| http://lotka.stanford.edu | Microsat | Several genetic distances and bootstrap procedures |
| ftp.cefe.cnrs-mop.fr under the directory pub/pc/msdos/genepop | Genepop | Test for Hardy–Weinberg, genotypic disequilibrium and population differentiation. Estimate $\rho_{ST}$ and $F_{ST}$, allele frequencies and Nm estimates. |
| http://anthropologie.unige.ch/arlequin/ | Arlequin | Several intra- and inter-population methods to describe variability and genetic differentiation; $\rho_{ST}$ is also calculated. |

# 7.4 POPULATION SIZES AND GENE FLOW

As mentioned above, some demographic events may disturb the interpretation of the results of genetic distance measures. Small genetic distances do not always hint at small divergence times between two populations. Large amounts of gene flow and/or constantly large population sizes prevent differentiation by genetic drift. Conversely, large genetic distance values may also reflect isolation and very small population sizes, and not necessarily long divergence times. Since we do not know the past of the populations, careful analyses of the results are mandatory. The degree of variability, measured by the gene diversity, may be very useful to evaluate differences in population sizes. Gene diversity or heterozygosity (H) may be estimated:

$$H = 2n(1 - \Sigma x_i^2)/(2n - 1),$$

where $n$ is the sample size and $x_i^2$ the gene frequency of the ith allele. This estimate is unbiased and appropriate for small populations (Nei & Roychoudhury, 1974).

Another informative indicator of genetic variability is the number of alleles. The larger the population size the greater is the number of alleles (k). A useful estimator (M) may be obtained from the expectation of H and k (Ewens, 1972). Under neutrality and genetic drift/mutation equilibrium $M = 4N\mu$, where $\mu$ is the mutation rate. Thus we obtain:

$$H = M/(M + 1)$$

$$k = 1 + [M/(M + 1)] + [M/(M + 2)] + [M/(M + 3)] + [M/(M + 4)] + ... + [M/(M + 2n - 1)]$$

The comparison of M obtained from, both, H and k ($M_H$ and $M_k$, respectively), shows hidden substructures of a population (Chakraborty *et al.*, 1997). Amalgamation of populations causes an excess of rare alleles, inflating $M_k$ (Chakraborty *et al.*, 1988). Significant differences between $M_H$ and $M_k$ may also reflect high gene flow. If we imagine that a population investigated receives significant numbers of migrants, a structure of an amalgamated population will be simulated.

Under genetic drift–mutation equilibrium the number of migrants per generation (Nm) may be accessed from $F_{ST}$ estimates (for a review see Nei, 1987; Slatkin & Barton, 1989). Therefore, we obtain $F_{ST}=1/(4Nm + 1)$. However, microsatellite loci follow the SMM and a more appropriate measure of gene flow from *F*–statistics seems to be obtained from $R_{ST}$ in populations at equilibrium (Slatkin, 1995; Santos *et al.*, 1997):

$$Nm = [(d - 1)/4d] [(1/R_{ST}) - 1],$$

where d is the number of populations sampled.

As mentioned above, the assumption of equilibrium is essential for some estimates. The term 'genetic drift–mutation equilibrium' denotes a situation where two opposing forces are in equilibrium, as in a chemical reaction. One force, genetic drift, tends randomly to eliminate alleles of populations. The other force, mutation, introduces new alleles into a population. The intensity of genetic drift is proportional to the population size. Thus, effective population size and mutation rate determine the dynamics of the equilibrium process. Gene flow also plays a role as an opposing force to genetic drift. To test whether populations are in genetic drift–mutation equilibrium, Slatkin (1993) suggested investigating a pattern of isolation by distance. This may be easily done by testing the correlation between genetic and geographic distance matrices using the Mantel test (Mantel, 1967). $F_{ST}$ and $R_{ST}$ estimates should be used.

In conclusion, several genetic distance measures are available, considering the different features of microsatellite loci, but the mode of evolution of these loci is far from being homogenous. Hence tests to access correlations between matrices of different genetic distance measures (Mantel test) can be suggested as an alternative to evaluate the consistency of the results obtained. However, gene flow and population size differences should not be underrated.

## 7.5   TANDEM REPEAT BLOCK EXPANSION DISEASES – A CONTINUUM FROM TRINUCLEOTIDES TO MINISATELLITES? IMPLICATIONS FOR DATABASE USAGE IN POPULATION GENETICS

Trinucleotide expansion diseases have recently attracted a great deal of attention, as, for example, model disorders for Mendelian late onset and/or neurodegenerative diseases in general. Therefore, suddenly, the properties of simple trinucleotide repeats generated increased interest. $(CAG)_n$ expansions are responsible for the class I trinucleotide block expansion (TBE) diseases, whereas class II TBE comprises enlarged $(CCG)_n$, $(CTG)_n$ and $(GAA)_n$ tracts (Paulsen & Fischbeck, 1996). The respective $(CAG)_n$ blocks in class I TBE disorders are situated in the exons of apparently completely unrelated genes and they are translated into perfect polyglutamine tracts (Gutekunst *et al.*, 1995; Trottier *et al.*, 1995). In the class II TBE the pathogenetic repeat expansions are located in introns, 5' untranslated regions (UTRs) or 3' UTRs, thus inhibiting efficient gene expression (see e.g. Campuzano *et al.*, 1996). Previously, fragile chromosomal sites were detected (e.g. on the human X chromosome, FRAX). Meanwhile, these cytogenetic peculiarities have been identified as tremendously elongated blocks of simple $(CCG)_n/(CGG)_n$ repeats (Verkerk *et al.*, 1991). In contrast, oculopharyngeal muscular dystrophy is caused by comparatively short $(GCG)_n$ expansions in the gene coding for the poly(A) binding protein (Brais *et al.*, 1998).

In synpolydactyly (SPD) a cryptic simple repeat (GCG, GCA, GCT and GCC; Muragaki *et al.*, 1996) appears to be causative for triggering of the disease. The mechanism bringing about the expansion is likely unequal crossing over (Warren, 1997). In addition, there are trinucleotide blocks in the human genome that have the ability to expand in the normal population and they do not cause disease (Nakamoto *et al.*, 1997). Recently, several 'minisatellite diseases' have been clarified, such as another fragile chromosome site, FRA16B, where the expansion of a $(ATATATTATATATTATATCTAATATAT^C/_ATA)_n$ sequence is pathogenic (Yu *et al.*, 1997). Furthermore in progressive myoclonus epilepsy type 1 (Unverricht–Lundborg disease) the elongation of the minisatellite sequence $(CCCGCCCGCG)_n$ is responsible (Lafrenière *et al.*, 1997). Why is it that simple dinucleotide repeats apparently do not appear to over-expand like trinucleotides (and multiples thereof) and certain minisatellites to cause repeat block expansion diseases? This question cannot be answered completely convincingly at present but DNA structure is certainly part of the mechanism and the underlying cause for expansion (McMurray, 1995). In conclusion, neither the sequence content nor the repeat unit length of the pathogenic repeat (above a certain threshold) matters for the disease-causing repeat. Instead, the respective locations of these expansions are most crucial.

## 7.6  SUMMARY

At least several trinucleotide repeats in microsatellites and in certain mini-satellites are not in all cases ideal markers for analyses in population genetics. Depending on the exact location with respect to the neighbouring genes, these repetitive units could do much more harm than would be expected from a neutral marker suitable for population genetics. In addition, the relative mutation rates differ at di-, tri- and tetranucleotide microsatellite loci (Chakraborty *et al.*, 1997). Disease-causing trinucleotide blocks mutate up to seven-fold more often than tetranucleotide microsatellites (*op. cit.*). Finally, trinucleotide loci associated with diseases are under selective range constraints, introducing severe bias into the population genetic analysis. Such information should also be included in present data bases in order to avoid uncritical interpretations of the accumulated data. There are already sufficient indications that one has to be very cautious about the supposedly complete neutrality of all dinucleotide and minisatellite loci (Epplen *et al.*, 1998).

## 7.7  REFERENCES

Amos W., Sawcer S.J., Feakes R.W. & Rubinsztein D.C. (1996) *Nature Genet.* **13**, 390–391.

Boeke J.D. (1997) *Nature Genet.* **16**, 6–7.

Brais B., Bouchard J.-P., Xiel Y.-G., Rochefort D.L., Chrétien N., Tomé F.M.S, Lafrenière R.G., Rommens J.M., Uyama E., Nohira O., Blumen S., Korcyn A.D., Heutink P., Mathieu J., Duranceau A., Codère F., Fardeau M. & Rouleau, G.A. (1998) *Nature Genet.* **18**, 164–167.

Britten, R.J. & Kohne, D.E. (1968) *Science* **161**, 529–540.

Campuzano V., Montermini L., Moltò M.D., Pianese L., Cossée M., Cavalcanti F., Monros E., Rodius F., Duclos F., Monticelli A., Zara F., Canizares J., Koutnikova

H., Bidichandani S.I., Gellera C., Brice A., Trouillas P., De Michele G., Filla A., De Frutos R., Palau F., Patel P.I., Di Donato Mandel J.-L., Cocozza S., Koenig M. & Pandolfo M. (1996) *Science* **271**, 1423–1427.

Chakraborty R., Smouse P. & Neel J.V. (1988) *Am. J. Hum. Genet.* **43**, 709–725.

Chakraborty R., Kimmel M., Stivers D.N., Davison L.J. & Deka R. (1997). *Proc. Natl. Acad. Sci. USA* **94**, 1041–1046.

Cooper D.N. & Krawczak M. (1989) *Hum. Genet.* **83**, 181–188.

Deka R., Jin L., Shriver M.D., Yu L.M., Saha N., Barrantes R., Chakraborty R. & Ferrell R.E. (1996) *Genome Res.* **6**, 1177–1184.

Di Rienzo A., Peterson A.C., Garza J.C., Valdes A.M., Slatkin M. & Freimer N.B. (1994) *Proc. Natl. Acad. Sci. USA* **91**, 3166–3170.

Epplen C., Epplen J.T., Frank G., Miterski B., Santos E.J.M. & Schöls L. (1997) *Hum. Genet.* **99**, 399–406.

Epplen J.T., Mäueler W. & Santos E.J.M. (1998) *Cytogenet. Cell Genet.* **80**, 75–82.

Ewens W.J. (1972) *Theor. Popul. Biol.* **3**, 87–112.

Freimer N.B. & Slatkin M. (1997) *Microsatellites: evolution and mutational processes. Variation in human genome*, (*Ciba Foundation Symposium*). Wiley, Chichester, pp. 51–72.

Goldstein D.B. & Pollock D.D. (1997) http://lotka.stanford.edu/distance

Goldstein D.B., Ruiz Linares A., Cavalli-Sforza L.L. & Feldman M.W. (1995) *Genetics* **139**, 463–471.

Gutekunst C.-A., Levey A.I., Heilman C.J., Whaley W.L., Yi H., Nash N.R., Rees H.D., Madden J.J. & Hersch S.M. (1995) *Proc. Natl. Acad. Sci. USA* **92**, 8710–8714.

Jabbari K., Caccio S., Pais de Barros J.P., Desgres J. & Bernardi G. (1997) *Gene* **205**, 109–118.

Jeffreys A.J., Wilson V. & Thein S.L. (1985a) *Nature* **314**, 67–73.

Jeffreys A.J., Wilson V. & Thein S.L. (1985b) *Nature* **314**, 76–79.

Lafrenière R.G., Rochefort D.L., Chretien N., Rommens J.M., Cochius J.I., Kalviainen R., Nousiainen U., Patry G., Farrell K., Soderfeldt B., Federico A., Hale B.R., Cossio O.H., Sorensen T., Pouliot M.A., Kmiec T., Uldall P., Janszky J., Pranzatelli M.R., Andermann F., Andermann E. & Rouleau G.A. (1997) *Nature Genet.* **15**, 298–302.

Levinson G. & Gutman G.A. (1987) *Nucl. Acids Res.* **15**, 5323–5338.

Litt M. & Luty J.A. (1989) *Am. J. Hum. Genet.* **44**, 397–401.

Mantel N. (1967) *Cancer Res.* **27**, 209–220.

McMurray C.T. (1995) *Chromosoma* **104**, 2–13.

Michalakis Y. & Excoffier L. (1996) *Genetics* **142**, 1061–1064.

Muragaki Y., Mundlos S., Upton J. & Olsen B.R. (1996) *Science* **272**, 548–551.

Nakamoto M., Takebayashi H., Kawaguchi Y., Narumiya S., Taniwaki M., Nakamura Y., Ishikawa Y., Akiguchi I., Kimura J. & Kakizuka A. (1997) *Nature Genet.* **17**, 385–386.

Nei M. (1987) *Molecular Evolutionary Genetics*. Columbia University Press, New York.

Nei M. & Roychoudhury (1974) *Genetics* **76**, 379–390.

Ohno S. (1972) Evolutional reason for having so much junk DNA. In *Modern Aspects of Cytogenetics: Constitutive Heterochromatin in Man*, Pfeiffer RA (ed.) F.K. Schattauer, Stuttgart, pp. 169–180.

Ohno S. & Epplen J.T. (1983) *Proc. Natl. Acad. Sci. USA* **80**, 3391–3395.

Paulsen H.L. & Fischbeck K.H. (1996) *Annu. Rev. Neurosci.* **19**, 79–107.

Primmer C.R., Ellegren H., Saino N. & Moller A.P. (1996) *Nature Genet.* 13, 391–393.

Rousset F. (1996) *Genetics* 142, 1357–1362.

Royle N.J., Clarkson R.E., Wong Z. & Jeffreys A.J. (1988) *Genomics* 3, 352–360.

Santos E.J.M., Epplen J.T. & Epplen C. (1997) *Hum. Heredity* 47, 165–172.

Santos E.J.M., Epplen J.T., Guerreiro J.F. & Epplen C. (1998) *Hum. Hered.* 13, 57–64.

Schloetterer C. & Tautz D. (1992) *Nucleic Acids Res.* 20, 211–215.

Shriver M.D., Jin L., Ferrell R.E. & Deka R. (1997) *Genome Res.* 7, 586–591.

Slatkin M. (1993) *Evolution 1993* 47, 264–279.

Slatkin M. (1995) *Genetics* 139, 457–462.

Slatkin M. & Barton N.H. (1989) *Evolution* 43, 1349–1368.

Takezaki N. & Nei M. (1996) *Genetics* 144, 389–399.

Tautz D. (1989) *Nucl. Acids Res.* 17, 6463–6471.

Tautz D. (1993) In *DNA Fingerprinting: State of the Science*, Pena S.D.J., Chakraborti R., Epplen J.T. and Jeffrey, A.J. (eds). Birkhäuser, Basel, pp. 21–28.

Tautz D., Trick M. & Dover G.A. (1986) *Nature* 322, 652–656.

Trottier Y., Devys D., Imbert G., Saudou F., An I., Lutz Y., Weber C., Agid Y., Hirsch E.C. & Mandel J.-L. (1995) *Nature Genet.* 10, 104–110.

Valdes A.M., Slatkin M. & Freimer N.B. (1993) *Genetics* 133, 737–749.

Verkerk A.J.M.H., Pieretti M., Sutcliffe J.S., Fu Y.-H., Kuhl D.P.A., Pizzuti A., Reiner O., Richards S., Victoria M.F., Zhang F., Eussen B.E., van Ommen G.-J.B., Blonden L.A.J., Riggins G.J., Chastain J.L., Kunst C.B., Galjaard H., Caskey C.T., Nelson D.L., Oostra B.A. & Warren S.T. (1991) *Cell* 65, 905–914.

Warren S.T. (1997) *Science* 275, 408–409.

Weber J. & May P.E. (1989) *Am. J. Hum. Genet.* 44, 388–396.

Weber J. & Wong C. (1993) *Hum. Mol. Genet.* 2, 1123–1128.

Weir B.S. & Cockerham C.C. (1984) *Evolution* 38, 1358–1370.

Yandava C.N., Gastier J.M., Pulido J.C., Brody T., Sheffield V., Murray J., Buetow K. & Duyk G.M. (1997) *Genome Res.* 7, 716–724.

Yu S., Mangelsdorf M., Hewett D., Hobson L., Baker E., Eyre H.J., Lapsys N., Le Paslier D., Doggett N.A., Sutherland G.R. & Richards R.I. (1997) *Cell* 88, 367–374.

Zhivotovsky L.A. & Feldman M.W. (1995) *Proc. Natl. Acad. Sci. USA* 92, 11549–11552.

# 8    Gene Feature Identification

*Phillip Bucher*

Institut de Suisse Recherches Experimentales sur le Cancer, Vaud, Switzerland

## 8.1   INTRODUCTION

The ongoing genome sequencing projects are producing large quantities of raw sequence data. What can automatic sequence analysis methods tell us about the function of these sequences? Several chapters of this book address this general question. This chapter introduces gene feature prediction methods for the nonprotein-coding part of the genome, including tools to detect RNA genes. The emphasis is on higher eukaryotic genomes, although certain problems in microbial sequence analysis will also be addressed. The goal is to give the non-expert reader a basic understanding of gene feature identification methods, and to enable him to apply these tools in his own research.

In the beginning of the genome era, development of automatic function prediction methods focused on protein sequences and on method to locate protein-coding regions in nucleic acid sequences. Regulatory genomic regions received less attention by computational biologists and, as a consequence, the tools to analyse them are less advanced. The reasons for this past trend are manifold. One reason was that homology-based function prediction methods are more effective for protein sequences than for DNA sequences because the protein-coding regions are usually conserved over longer evolutionary periods than the noncoding parts of the genomes. From an applied biomedical perspective, proteins are more relevant than noncoding DNA because they constitute potential therapeutic agents and drug targets. As a result, recent private and public sequencing efforts have focused on cDNA rather than genomic sequences.

With the expected rapid progress of the human genome project in the near future, the emphasis in sequence-based function prediction will shift to the regulatory regions of the genomes. Obviously, in order to understand a multicellular organism, it is not only necessary to know what the genes are doing, but

*Genetics Databases*
ISBN 0-12-101625-0

also when, where, and under which environmental conditions they are expressed. There will be two major challenges in this area: first to distinguish sequences containing relevant genetic information from the presumably much larger part of function-less DNA, and second to predict gene expression patterns and regulatory features from sequence. Considerable progress has been made with regard to the first problem, mainly due to our understanding of the evolutionary origin of repetitive DNA. However, the tools for predicting regulatory DNA elements are, with a few exceptions, still not yet reliable and selective enough for large-scale automatic genome annotation, although significant progress has been made recently. These tools must therefore be applied with great caution.

The remaining part of this chapter is organized as follows: Section 2 provides an overview of sequence features covered along with biological background information, Section 3 introduces the basic sequence analysis methods used to identify these features, Section 4 contains a list of publicly available tools implementing these methods, and Section 5 presents an example showing the application of these tools to a mammalian gene regulatory region.

## 8.2  BIOLOGICALLY INTERESTING SEQUENCE FEATURES

What are biologically interesting sequence features of noncoding sequences? The answer to this question given here may be partly subjective as it inevitably reflects the author's research interests and experience. This chapter primarily covers sequence features that are interesting from a gene functional and evolutionary perspective. Some of them, e.g. repetitive elements, are also of practical importance as they can interfere with experimental procedures such as nucleic acid hybridization experiments. However, sequence features which are interesting primarily for technical reasons, such as restriction endonuclease cleavage sites, are not considered.

When interpreting computer-generated sequence annotations, it is important to understand how a particular feature was originally defined. At this conceptual level, three different cases can be distinguished.

- *Sequence-defined features.* Such features are exclusively defined by information contained in the sequence itself and thus are not subject to experimental verification. Examples are base compositional properties, simple repeats, or mathematically defined sequence complexity measures.
- *Predictions of physical, chemical, or biological properties.* These features are ultimately defined by a physical, chemical, or biological

assay. The automatically generated sequence annotations thus represent testable hypotheses. Examples of such features are DNA curvatures, protein-binding sites, and promoters.

- *Sequence homology.* This is an evolutionary sequence property. Sequence homology identified by a computer method cannot be falsified by an experiment but can be shown to be inconsistent with a model of molecular evolution. In this sense, a claim about homology is falsifiable. The Alu family of human repetitive elements exemplifies a sequence class defined by homology.

A selection of sequence features will now be introduced that can be identified by existing software tools. Note that the order in which they are presented is based on biological considerations and thus does not strictly follow the above classification.

## 8.2.1 Compositional features: isochores, CpG islands

The compositional properties of DNA sequences discussed here include DNA base composition and short oligomer (word) distributions up to the hexanucleotide level. In information theory, the characterization of these properties is known as Markov chain analysis. Significant differences in base and oligonucleotide compositions have been observed between organisms as well as between function classes of DNA sequences. The analysis of these properties can therefore give important indications about the origin and physiology of a particular sequence. Based on oligonucleotide frequency distributions it was, for example, proposed that the variable sequences of yeast mitochondrial genomes arose from recent horizontal gene transfers (Pietrokovski & Trifonov, 1992). Compositional properties have also been used in gene prediction to distinguish introns from coding exons (for a recent evaluation of these methods, see Guigó & Fickett, 1995). Two examples of compositional features which are particularly important to the understanding of vertebrate genomes, will be described in more detail.

### 8.2.1.1 Isochores

In many eukaryotes, especially vertebrates, chromosomal regions of high G+C content alternate with regions of lower G+C content. Within such regions, which have been termed 'isochores' (for a recent review, see Bernardi, 1995), the base composition is remarkably stable although characteristic differences between function classes (e.g. coding versus noncoding sequences) are maintained. In the human genome, five isochore classes have been distinguished, which are designated (in order of increasing G+C content) L1, L2, H1, H2, and H3. The average size of an isochore is in the range of 300 kb. The assignment of a human gene to a particular isochore class is of interest for several reasons. Since gene density appears to increase with G+C content, G+C rich isochores are usually gene-rich and genes therein tend to be compact by having short introns. It is noteworthy in this context that isochore assignments can be made from mRNA sequences as well, most reliably by taking only the third codon position into account. It has further been proposed that different isochore classes replicate at different times during the cell cycle, possibly related to changes in the deoxyribonucleotide pools. Knowledge of the regional G+C content is also important for assessing the statistical significance of other sequence features, for example transcription control signals. As an example, a predicted TATA-box promoter element is less likely to occur by chance in a G+C rich isochore than in an A+T rich isochore, and thus is more likely to be biologically significant if it occurs in such an environment.

### 8.2.1.2    CpG islands

The CpG dinucleotide is about three times under-represented in vertebrate genomes. This phenomenon has been attributed to the fact that a large proportion of CpG dinucleotides are methylated at the C residue, a DNA modification not found in other dinucleotide contexts. It was then speculated that methylated C more frequently mutates into T than nonmethylated C, leading to a CpG depletion of the genome. The CpG repression is however not uniform. In the genomes of warm-blooded vertebrates there are so-called CpG islands (Gardiner-Garden & Frommer, 1987), which were originally defined as regions of 200–2000 bp showing three characteristic properties: (1) absence or only weak under-representation of CpG dinucleotides, (2) undermethylation of CpG residues, and (3) a G+C content of 50% or higher. The second criterion is of little practical importance because the methylation status of a C residue in a DNA sequence is usually not known. The identification of CpG islands is of interest because they often overlap with promoter regions, the centre position typically located about 100 bp downstream from the initiation site of transcription. It was estimated that about 60% of human promoters are associated with CpG islands (Antequera & Bird, 1993). The reasons for this co-localization are not entirely clear. One popular hypothesis assumes that the under-methylation of CpG is the consequence rather than the cause of physiological processes leading to transcription initiation in germ-line cells. This would also explain why CpG islands are predominantly found at the 5′ ends of housekeeping genes.

## 8.2.2 Repeats: microsatellites, minisatellites, transposable elements

The genomes of higher organisms contain many types of repeats (see Chapter 7 and Heringa 1998) belonging to two main classes: tandem repeats and interspersed repeats. The tandem repeats are further classified into microsatellites, minisatellites, and tandemly repeated gene families. Microsatellites are characterized by very short repeat units, mainly mono-, di-, and tri-nucleotides. Minisatellites have somewhat longer repeat units (up to about 100 bp). The third class includes genes for histone proteins and ribosomal RNA that, in certain species, exist as large tandemly repeated gene families. Clusters of micro- and mini-satellites occur in the telomere and centromere regions of the chromosome. The term 'satellite DNA' was first applied to the sequences in the centromeres of mouse chromosomes which could be isolated as a minor fraction of the genome that differed in buoyant density from the rest of the DNA. Note that the modern definition of the term has considerably diverged from its original meaning.

While minisatellites and tandemly repeated gene families are relatively rare, microsatellites are extremely frequent and occur everywhere in the genome. Identification of microsatellites is important for two reasons. First, most microsatellites are polymorphic and thus provide useful molecular markers for population studies, genetic map construction, and positional cloning of genes corresponding to a known phenotype. The polymorphic nature of microsatellite loci is commonly explained by a DNA polymerase slippage mechanism leading to a change in copy number of oligonucleotide repeat units. Second, expansion of trinucleotide repeats appears to be a frequent cause of human genetic disease. The observation of such repeats, especially in coding regions, is therefore an indication that a gene could be particularly susceptible to deleterious mutation.

Interspersed repeats represent active or inactive (fossil) copies of transposable elements of various classes (Smit, 1996). The large elements, up to several kb long, are sometimes referred to as LINEs (long interspersed elements) and the short ones (below 500 bp) as SINEs (short interspersed elements). Note that most transposable elements do not really move from one place in the genome to another but rather create additional copies of their own sequences at new places. Based on transposition mechanisms one distinguishes between DNA transposons and retrotransposons. DNA transposons insert themselves into new sequences by a recombination process mediated by an integrase, which is typically encoded by the transposon itself. Retrotransposons spread through the genome via a copy mechanism resembling the replication cycle of retroviruses. Large retrotransposons typically encode their own reverse transcriptase whereas small retrotransposons depend on *trans*-acting DNA polymerase genes from other sources. SINEs are thought to originate from RNA genes with internal POL III promoters, primarily tRNA and 7SL RNA genes. The presence of an internal promoter enables these elements to be transcribed after insertion at a new place independently of flanking sequences.

SINEs and LINEs are viewed by many as parasitic or selfish DNA elements encoding gene functions beneficial only to their own replication cycle and possibly deleterious to the host genome. Many naturally occurring mutations including several that are responsible for human genetic diseases are insertions of repetitive elements. An alternative view is that the large number of repeats contributes to the evolutionary plasticity of a genome by allowing exon shuffling through nonhomologous recombination events. Parasitic expansion of particular LINE or SINE families must have happened many times during the evolutionary history of higher organisms. The large majority of copies found in present genomes are fossils – sequences that have been inserted some time ago and since then have accumulated mutations that prevent them from generating new copies via a full transposition cycle. The most frequent human repetitive elements, the Alu and LINE1 families, which together constitute about 20% of the human genome, result from recent and probably still ongoing expansions, and thus are considered young elements. In contrast, the MIRs (mammalian-wide interspersed repeats) are ancient elements which have spread through an ancestral mammalian genome more than 150 million years ago. Ancient elements differ from young elements by having a much higher degree of sequence diversity, and in that they are found at homologous positions in the genomes of different species. While two Alu elements typically show over 95% sequence identity among themselves, the similarity between MIRs is often barely detectable with the currently most sensitive sequence similarity search techniques.

The identification of SINEs and LINEs is an important part of the analysis of noncoding sequence regions for the mere reason that they are so frequent. Interspersed repeats are usually assumed to have no function and therefore indirectly may help to pinpoint sequence regions encoding structural genes or important regulatory information.

## 8.2.3   Evolutionarily conserved noncoding sequences

Cross-species comparisons of homologous sequences allows identification of regions that are better conserved than others and thus are more likely to serve an important function. Although the specific functions remain unknown in such cases, this method, which has been termed phylogenetic footprinting, constitutes a powerful strategy to select interesting sequence regions for further experimental characterization. In vertebrates, it is assumed that sequences evolving under no selective constraints undergo complete randomization in about 100 million years. 'Complete randomization' in this context means that no statistically significant sequence similarity can be detected. Detectable sequence conservation between a mammalian and a nonmammalian species thus constitutes evidence for biological function. A comprehensive analysis of homologous noncoding sequences from vertebrate genomes has revealed many highly conserved regions (HCRs) in 3' untranslated regions of mRNAs (Duret *et al.*, 1993). Some of these HCRs were shown to be involved in translational regulation and others in mRNA targeting to intracellular compartments. Phylogenetic footprinting has also proven to be a powerful technique to identify important transcriptional control regions in the 5' flanking regions of genes, as exemplified by the case study presented in Section 5.

## 8.2.4    DNA regulatory elements: promoters, transcription factor binding sites, complex regulatory regions

In eukaryotes, transcriptional regulation is thought to be mediated by short elements which bind to gene regulatory proteins called transcription factors. These elements were often found not to function when transferred into a different sequence context. The emerging current view is that autonomous control regions typically consist of multiple modules of the same or different type and that the order, orientation, and spacing between the components is critical for function.

Individual elements can either be defined biochemically as protein-binding sites, or genetically as sequence regions producing some observable effects when mutated. It is important to recognize that for a given element, these two alternative definitions may not exactly overlap, and that the physiological relationship between a genetic concept (e.g. a promoter element) and a biochemical concept (e.g. a transcription factor binding site) is generally very difficult to establish. For this and other reasons, the classification and nomenclature of transcriptional control elements remains incoherent and controversial. The still modest performance of the current tools to identify such elements results partly from the fact that the targets are ill-defined.

Composite regions have been classified into 'promoters' located near the start sites of transcription, 'enhancers' acting from remote locations, 'locus control regions' coordinately regulating the expression of a gene cluster, 'matrix attachment sites' (MARs), and others. Again, the terminology is not very coherent. For example, sequence regions as close as 100 bp upstream of the transcription start site have been presented as enhancers, while others located up to 300 bp upstream have been considered integral parts of the promoter region. From a biochemical or physiological viewpoint, the distinction between enhancers and promoters may not be justified since the same classes of elements and similar regulatory effects were found to be associated with both types of regulatory regions.

In the biological literature, the term promoter is not only used in the sense of a transcription initiation control region, it sometimes simply refers to a transcription start site. For example, the Eukaryotic Promoter Database (Cavin Perier *et al.*, 1998), which has been used in the development of most promoter prediction algorithms, is a collection of experimentally defined transcription initiation sites rather than a database of gene control elements. As a consequence, eukaryotic promoter prediction programmes function as initiation site prediction tools (Fickett & Hatzigeorgiou, 1997).

## 8.2.5   RNA genes and RNA regulatory elements

About 3% of all human genes may encode RNAs. Although they are coding regions in a literal sense, they are quite different from protein-coding sequences from the sequence analyser's perspective and therefore are considered in this chapter. Certain genes, for example ribosomal RNA genes are strongly conserved and thus readily identified by sequence similarity searches. The less conserved ones, including tRNA genes, require specialized computer programs for their identification and therefore are easily overlooked. Note that RNA genes have been found everywhere in the genome, also in introns of other genes.

A number of regulatory elements have been identified within the noncoding part of mRNAs. Some of them were shown to be target sites of sequence-specific RNA-binding proteins. A classical example is the iron-responsive element (IRE) bound by iron-regulatory proteins (Henderson, 1996). Other regulatory RNA signals may interact with complementary RNAs, for instance nuclear anti-sense RNA (Lipman, 1997). The computational problem of identifying these elements is analogous to the problem of recognizing transcriptional control elements. However, more sophisticated algorithms are required to capture the conserved features of functional RNA sequences if secondary structure is involved in the biological recognition process.

## 8.2.6   Sequence features in prokaryotic and organelle genomes

Many of the sequence features introduced so far are also relevant to prokaryotic and organelle (mitochondrial and chloroplast) genomes. For example, interspersed repeats and transposons also exist in bacteria but make up a much smaller proportion of the total DNA. Gene regulatory elements can be identified with the same computational tools. However, there are also a number of characteristic differences between eukaryotic and prokaryotic genome organization. Among the additional features that are of interest in prokaryotic and organelle genomes are self-splicing introns and inteins. The latter are intervening protein sequences that are spliced out from the primary translation product. Prokaryotic transcription initiation mechanisms differ from eukaryotic mechanisms in that the RNA polymerase directly recognizes the sequence upstream from the initiation site. Prokaryotic promoter prediction is thus viewed as a protein-binding site rather than a transcription initiation site prediction problem. Moreover, prokaryotic genes are organized in operons that are co-transcribed into polycistronic mRNAs. The identification of operon boundaries can be helpful in predicting regulatory properties of genes, and thus represents an important objective in prokaryotic sequence feature prediction, with no obvious counterpart in vertebrate genome analysis. Another peculiar phenomenon of bacterial genetics is the frequent exchange of DNA between organisms that are not closely related. The identification of genomic regions with atypical compositional properties helps to identify genes which may have been acquired recently through such horizontal gene transfer.

From a purely practical perspective, sequence feature identification in bacterial genomes appears to be more difficult because fewer tools are available to biologists. Until recently, virtually all prediction algorithms were developed only for *Escherichia coli*. If the current trend of sequencing several bacterial genomes per year continues, more and better tools can be expected to become available soon.

## 8.2.7   DNA structural properties

A certain number of DNA structural properties have been implicated in physiological processes such as transcription, replication, recombination, and repair. Also, from a theoretical perspective, it seems plausible that these sequence-dependent DNA features would interfere with metabolic processes of genes. Among the most commonly analysed physical parameters of DNA are thermostability, Z-DNA forming potential, intrinsic DNA-curvature and bendability. Several computational methods have been developed to predict these parameters from sequence. Although such features can provide some clues about the function of a gene, they must be interpreted with great caution, as the relationship between predicted physical and experimentally shown biological properties of DNA is poorly understood.

The nucleosome positioning problem may be viewed as an extension of a DNA structure prediction problem. In the nucleus of a eukaryotic cell, DNA is not active in naked form but functions in a nucleoprotein complex called chromatin. The basic building block of chromatin is the nucleosome, consisting of a 145 bp segment of DNA wrapped around a histone octamer. Between nucleosomes, there is a short linker region where the DNA is more accessible to soluble proteins, including transcription factors. Not surprisingly, the precise positioning of nucleosome was found to play a role in the transcriptional regulation of several genes. Knowing the exact centre position of a DNA-histone octamer complex may be important for the understanding of gene regulatory processes because it allows one to predict which minor or major groove sequence features of a DNA region are exposed to the solvent and thus accessible to transcription factors (for a tutorial review, see Trifonov, 1991).

## 8.3    SEQUENCE ANALYSIS METHODS

The existing tools for gene feature prediction in noncoding sequences rely on a great variety of methodological concepts. Some of them, for example sequence comparison algorithms, are widely applied and fairly well known by most biologists. The description of these methods focuses on essential aspects relevant to the features covered by this chapter. Important concepts such as weight matrices, which were designed specifically for the characterization of gene control elements, will be explained in more detail. For an introduction to more specialized or advanced methods, the reader will be referred to other publications.

### 8.3.1    Sequence comparison and similarity search methods

There are two major classes of sequence comparison methods: alignment algorithms and dot matrix methods. A sequence alignment algorithm produces two types of output: a similarity score and the alignment itself. The similarity score is useful, for example, for assessing the statistical significance of an observed match, or for the estimation of an evolutionary distance. The alignment indicates which residue in one sequence corresponds to which residue in the other sequence. Such a mapping may help to locate by homology a physiological site in a new sequence, e.g. a transcription start site.

From a mathematical perspective, a sequence alignment algorithm solves an optimization problem, namely to find the best or the $n$ best alignments between two sequences. The quality of the alignment is measured by an alignment score which is usually equal to the sum of the weights of the matches, mismatches and gaps occurring in the alignment. For DNA comparisons, two different weights are used for matches and mismatches, respectively and gaps are scored by a linear length-dependent function.

Two types of sequence alignment algorithms are commonly distinguished, global and local. A global algorithm attempts to align two sequences over their entire lengths, regardless of whether there is any similarity at all in a particular region. A local alignment algorithm searches for matching segments between two sequences. For most applications to noncoding sequences, the second type of approach is more useful. Note that the alignment scoring system is critical in local similarity searches as it determines the threshold per cent of identity of the alignments found. For example, the combination of match weight 1 and mismatch weight –3, requires 75% sequence identity for extension of an alignment. If the similarity drops below this value, the matches will no longer compensate the negative contributions of the mismatches and the extended alignment would thus be worse than a shorter alignment in terms of alignment score.

Sequence alignment algorithms are often used to compare a query sequence against a large sequence database. The Smith–Waterman algorithm (Smith & Waterman, 1981) computes optimal alignment scores for all sequence pairs. Since it is a rigorous method guaranteed to find the optimal solution to the problem, it is computationally expensive and often deemed too slow for searching large databases. Note, however, that for searching one of the smaller specialized sequence databases listed in Table 8.2 (see Section 8.4), the search time is usually acceptable. To speed up database searches, faster heuristic algorithms have been developed such as FASTA (Pearson, 1990) and BLAST (Altschul *et al.*, 1990, 1997). These algorithms attempt to produce the same result as the Smith–Waterman algorithm, although by a procedure that is not guaranteed to find the optimal alignment. Therefore, they may occasionally miss a significant match.

In the case of weak sequence similarity, it is important to perform a test to assess whether the score of the alignment is significantly higher than what one would expect by chance. The *P*-values reported by the newer versions of most sequence similarity search programmes are not absolutely reliable. Cases have been reported, where BLAST *P*-values of $10^{-5}$ turned out not to be significant upon further tests. The classical procedure to evaluate alignment significance consists of repeatedly shuffling one of the two sequences and computing similarity scores for the shuffled sequences. The score distribution obtained this way is then fit to an extreme value distribution, which can be used to compute a *P*-value for the optimal alignment score obtained with the real sequences (for more details, see Pearson, 1990).

A dot matrix is a rectangular graphic representation of local similarity between two sequences (an example is shown in Fig. 8.1). One sequence corresponds to the horizontal, the other to the vertical axis. The intensity of a dot indicates the degree of similarity of two short sequence segments of fixed lengths at particular coordinates. Since no gaps are introduced to maximize the number of residue matches between the sequence segments, dot matrix methods are slightly less sensitive than alignment methods. However, they show the whole picture of local sequence similarity relationships between two sequences at once.

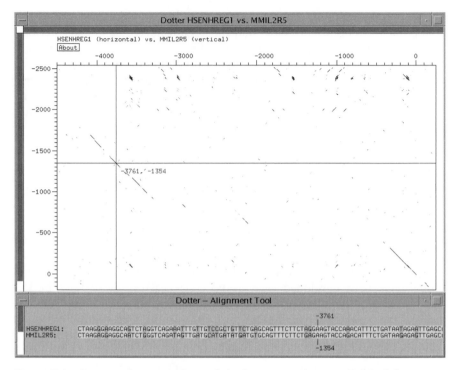

**Figure 8.1**    Dot matrix comparison of the human and mouse IL-2R alpha promoter regions. The alignment window in the lower part of the figure shows a part of the left-most highly conserved sequence region indicated in Fig. 8.2, which contains the IL-2 responsive enhancer found by experiments.

## 8.3.2    Consensus sequences and weight matrices

The target sites of transcription factors are often highly degenerate. In this context, the term degeneracy means that different sequences can bind the same factor. However, such sequences nevertheless exhibit similarity among themselves and seem to conform to some loose consensus. Such patterns have traditionally been described by so-called consensus sequences and weight matrices. A consensus sequence is a string of base symbols which may include ambiguous IUPAC-codes such as R for A or G. A natural target site often does not conform exactly to the consensus sequence and one therefore has to allow one or a few mismatches when searching a putative gene control region with a consensus sequence.

An implicit assumption of consensus sequence-based methods is that all mismatches are equally detrimental. This, however, has not been found to be the case for virtually all protein-binding sites where the effects of alternative base substitutions have been measured experimentally. Some mismatches are usually found to have only mild effects on *in vitro* binding or *in vivo* function while others prove to be fatal. For this reason, consensus sequences are generally poor predictors of transcription factor binding sites or gene control elements.

Weight matrices are more flexible descriptors of degenerate sequence motifs which can account for varying degrees of severity of base substitutions in a functional element. The first weight matrix-like method was applied to ribosome-binding sites in *E. coli* and was then called a Perceptron (Stormo *et al.*, 1982). Staden (1984) introduced the term weight matrix and presented the basic methodology illustrated in Fig. 8.2, which is still in use today. Berg and von Hippel (1987) provided a statistical mechanical interpretation of a weight matrix, retrospectively justifying the arithmetic conversions proposed by Staden (1984). According to this theory, weights should be proportional to the logarithms of observed base frequencies increased by a small number.

In biochemical terms, a weight matrix describes the nucleotide sequence preferences of a DNA-binding surface of a gene regulatory protein. Each column corresponds to a surface area which may be described as a base-pair acceptor site. According to Berg and von Hippel (1987), the position-specific weights represent binding energies of individual base pairs bound to the protein. The weight matrix model further assumes that individual base pairs contribute independently to the total binding energy of a DNA–protein complex. Although this hypothesis is certainly a simplification, it has been shown to be a reasonable approximation in a number of test cases, e.g. for the Mnt repressor (Stormo *et al.*, 1993).

A weight matrix is a quantitative sequence motif descriptor which assigns a score to any nucleotide sequence of corresponding length. As with consensus sequences, a threshold score needs to be defined when searching a gene regulatory region for a weight matrix-defined sequence motif. The magnitudes of the scores reported for individual sites are of interest because weight matrices are applied for two purposes: to identify new sites in new sequences, and to estimate the protein-binding strength or physiological activity of known sites. The score also provides a first indication of the biological significance of a predicted transcription factor (TF) binding site or control element. For this purpose, it is important to know how many matches above a certain score one would expect by chance in a sequence of a certain length. The expected number of matches above a given score per nucleotides scanned (sometimes referred to as a false-positive rate) can be computed analytically (Claverie & Audic, 1996) or empirically by searching a random sequence database for binding sites.

Unlike the similarities detected by an alignment algorithm between homologous sequences, a consensus sequence or weight matrix match almost never represents a statistically significant event by itself. Typical false-positive rates are in the range of 1/100 to 1/10 000 bp, which is roughly in agreement with

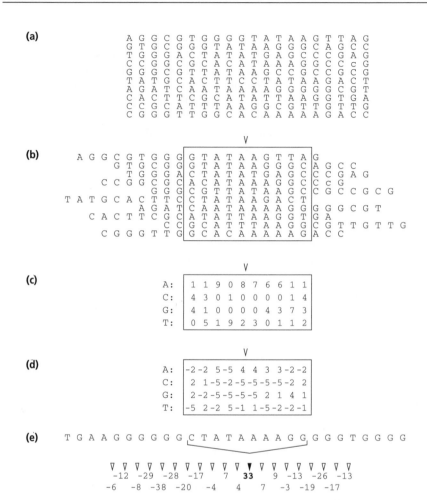

**Figure 8.2** Construction and usage of a weight matrix. (a) A set of sequences known to contain a certain feature (e.g. a transcription factor binding site) is selected as input data for construction of a computer-searchable weight matrix description. (b) A gap-free alignment is generated with a multiple alignment algorithm from the above sequence set. Note that this alignment optimizes local sequence similarity within a 10 bp region. (c) A base frequency table is derived from the aligned sequence regions. (d) A weight matrix is computed from the above base frequency table. The weights are log-odds ratios given by the following formula:

$$w = 10 \times \log10 \left[ (f+1)/(N/4+1) \right],$$

where $f$ is the observed base frequency at a particular position of the weight matrix and $N$ the total number of bases in the corresponding column of the sequence alignment. (e) The weight matrix is used for identifying the target feature in a new sequence. By sliding the weight matrix along the sequence, a score is computed for every 10 bp long subsequence. In the sequence example shown, there is a strong match with a score of 33. The other scores presumably fall below a reasonable cut-off value which may be in the range of 10.

the frequency estimates for *in vitro* binding sites to transcription factors based on biochemical data. For instance, the gene regulatory protein C/EBP recognizes on average one site in about 250 bp (Ryden & Beemon, 1989), implying a large excess of potential binding sites over protein molecules in the nucleus of a human cell. Therefore, most of the computer predicted or *in vitro*-identified C/EBP-binding sites are probably physiologically irrelevant. Additional criteria, such as the strength of the binding site, its conservation in a related organism, or its approximate co-localization with another sequence feature, will thus be necessary to assess the biological significance of a particular site.

In the preceding considerations, we have assumed that a weight matrix for a particular site already exists. But how are the parameters of such a matrix defined in the first place? There are two fundamentally different approaches to this problem:

- *Alignment of known binding sites*: the input data consists of a set of short DNA sequences known to bind to a particular protein or to have specific biological functions. The sequences may have been obtained by an *in vitro* selection–amplification experiment (Pollock & Treisman, 1990). Since the exact location of the binding sites within the sequences is unknown, the sequences first need to be aligned by a method that maximizes some consensus measure. A base frequency table is then derived from the relevant portion of the multiple sequence alignment and the position-specific base frequencies are subsequently converted into weights by some mathematical formula. Note that many different arithmetic conversion schemes have been proposed. For example, the matrices searched by the widely used MatInspector tool (Quandt *et al.*, 1995) are not computed by the previously mentioned formula introduced by Staden (1984).
- *Exhaustive mutagenesis*: a high-affinity binding site is selected as reference oligonucleotide. All single-base substitution mutants of the binding region are then generated and assayed *in vitro* for protein binding or *in vivo* for biological activity. The quantitative measurements obtained in this way are directly converted into weights according to some biophysical model. This rarely used approach is illustrated, for example, by the work of Kraus *et al.* (1996).

Weight matrices have mainly been applied to DNA signals but can also be used to characterize regulatory motifs in RNA sequences. The first weight matrix application described by Stormo *et al.* (1982) was for RNA-signal, translational start sites in *E. coli.* Many gene prediction algorithms use weight matrix-like modules for detecting intron–exon boundaries which, from a physiological viewpoint, are RNA processing signals. However, in other cases where RNA secondary structure is involved in the recognition process, a simple weight matrix will not be effective since the base-pairing rules applying to

RNA stemloop structures violate the basic assumption that base preferences at different positions in the motif are independent of each other.

## 8.3.3   Profiles and hidden Markov models

One limitation of weight matrix methods is that they do not allow for variable spacing between different parts of a binding site. However, many TF-binding sites consist of two conserved blocks separated by a variable spacer. For example, most homodimeric DNA-binding proteins have dyad-symmetrical binding sites consisting of two identical half sites arranged in opposite orientations. For some of them, it has been proposed that the spacing between the half sites could be of variable length. The best known example of an asymmetrical bipartite recognition sequence is the *E. coli* transcriptional promoter recognized directly by RNA polymerase.

Early approaches to search for such complex patterns have used combinations of weight matrices in conjunction with variable length linker scoring schemes (see for example the *E. coli* promoter prediction method implemented in program TargSearch; Mulligan *et al.*, 1984). The sequence profiles introduced by Gribskov *et al.* (1987) represent the first generalization of the weight matrix concept allowing for position-specific gap weights at any position. This profile type nevertheless has some limitations, most notably the impossibility to score insertion and deletion gaps separately, which renders its applications to variable length gene-control signals ineffective. The more recently introduced so-called generalized profiles provide more flexibility with regard to insertion and deletion scoring and are thus more adequate for this purpose. Indeed, it has been shown that the previously mentioned method for detecting *E. coli* promoters can be described by a generalized profile and emulated by corresponding search programmes (Bucher & Bairoch 1994). Note, however, that in contrast to protein family profiles, profiles characterizing DNA control elements make use of insertion and deletion opportunities in a very restrictive manner. The *E. coli* promoter profile, for example, allows the opening of a deletion gap only at a single position and does not permit insertions at all.

The more recently introduced hidden Markov models (HMMs) have become another widely used tool for characterizing degenerate sequence motifs (for a concise review, see Eddy, 1996). HMMs are not genuinely different from weight matrices or profiles: they represent an alternative formalism for describing the same kind of methods. It has been shown that models based on the standard architecture for molecular sequence families introduced by Krogh *et al.* (1994) can be converted into equivalent generalized profiles. The most obvious difference between the two descriptor types is that in the HMM representation, the individual parameters are probabilities which have to be multiplied for computation of a global score for a matching sequence region, whereas in a profile, the parameters are additive weights with no a priori specific interpretation attached to it. In contrast to profiles, HMMs are based on solid mathematical foundations, and their parameters can therefore be derived by well-established algorithms such as Expectation–Maximization (EM) or simulated annealing. For these reasons, the HMM formalism is preferable and expected gradually to replace the profile and weight matrix terminology in the future. However, at present, relatively few native HMM-based tools for the identification of specific gene features are available to biologists.

## 8.3.4   Neural nets

Neural nets constitute another family of methods which has been used for characterizing genetic control elements, e.g. eukaryotic promoter (Matis *et al.*, 1996). A useful introductory essay has been published by Hinton (1992). Very briefly, and in very general terms, a neural net is an information processing system that can learn from training data. After training, it can be used to distinguish certain objects from others, for example certain sequence elements from other sequences. Neural nets can also be trained to make quantitative predictions, for example about the strength of DNA–protein-binding site. The simplest type of neural net, the Perceptron, is equivalent to a weight matrix in its trained state. More complex nets have additional parameters residing in so-called hidden layers. Such nets have additional capabilities such as nonlinear response functions to input signals and mechanisms that can account for residue dependencies between different positions within a signal, which cannot be reproduced by HMMs or profiles. Conversely, neural nets have great difficulties in dealing with variable length motifs because the input layer is typically a rigid structure with a fixed number of cells, accepting sequences of only one length class. Another difference from the previously introduced methods is that neural nets are designed as black-box methods. While the weights of a weight matrix are usually known to the user and lend themselves to a physical interpretation, the parameters of a neural net are hidden and not meant to be biologically interpretable or of interest to the user. Thus, while neural nets may be very powerful function prediction tools, they usually do not tell us anything about the underlying molecular recognition process.

## 8.3.5   Methods for the analysis of RNA and DNA structural motifs

As mentioned earlier, the standard HMM and profile descriptors cannot deal with position interdependencies in sequence motifs, such as those occurring in

RNA secondary structures. More flexible methods exist which can deal with these problems. The extension of basic weight matrix methods to the dinucleotide level accounting for dependencies between adjacent bases (sometimes called weight array matrix, see Zhang & Marr, 1993), is straightforward. An early application of this approach to *E. coli* rho-independent terminators was described by Brendel and Trifonov (1984). Stochastic context-free grammars (SCFGs), a more flexible method related to HMMs, can serve as a general framework for describing RNA secondary structure motifs and formulating corresponding algorithms (Grate, 1995). It is beyond the scope of this chapter to introduce the concepts underlying these more advanced techniques.

Most methods to compute DNA structural features rely on physical and geometric parameters assigned to dinucleotide steps (Lavery & Sklenar, 1988; Shpigelman *et al.*, 1993). Algorithms to predict nucleosome positions also are based on a kind of dinucleotide sequence pattern (Ioshikhes *et al.*, 1996).

## 8.4 COMPUTER PROGRAMS, DATABASES AND WWW SERVERS

A large number of tools exist for gene feature identification. This Section briefly introduces a few programs, databases, and WWW servers which enable the reader to perform the various kinds of analyses described in the previous chapters. It should be noted that this selection remains incomplete and is partly subjective. Preference has been given to resources that are freely available to both academic and private research institutions, and to software which runs under UNIX operating systems. The internet addresses for accessing these tools can be found in Tables 8.1–8.3.

The most commonly used sequence similarity search program is BLAST (Altschul *et al.*, 1990, 1997) which exists in two versions, one from Washington University and one from the NCBI (National Center for Biotechnology Information). BLAST is very fast, but for nucleotide sequence searches not very sensitive owing to hard-coded limitations. FASTA (Pearson, 1990, 1994), an older, heuristic approximation of the Smith–Waterman algorithm, is generally slower, but offers more flexibility in balancing speed versus sensitivity. The FASTA package also includes a program that performs a rigorous Smith–Waterman local alignment search algorithm (ssearch3).

One of the most important applications of sequence similarity search in automatic genome annotation is the identification of repetitive elements. Two very useful databases exist for this purpose: Repbase (Jurka *et al.*, 1998), a compilation of interspersed repeats, and the 'simple sequences' collection (Jurka & Pethiyagoda, 1995), which contains frequently occurring human microsatellite types. Longer tandem repeats including minisatellites can be identified with programs such as 'satellites' (Sagot & Myers 1998) or the

**Table 8.1**  Selected computer programs useful for gene feature identification.

| Software package (program) | URL |
|---|---|
| **Sequence similarity search** | |
| NCBI-Blast2 | ftp://ncbi.nlm.nih.gov/blast/ |
| Wublast | http://blast.wustl.edu/ |
| FAST3(FASTA3, SEARCH3) | ftp://ftp.virginia.edu/pub/fasta/ |
| **Pairwise sequence comparison** | |
| FASTA2(lalign, prss) | ftp://ftp.virginia.edu/pub/fasta/ |
| Dotter | ftp://ncbi.nlm.nih.gov/pub/esr/dotter/ |
| **Repeats, low complexity regions** | |
| Dust | ftp://ncbi.nlm.nih.gov/pub/tatusov/dust/ |
| SEG | ftp://ncbi.nlm.nih.gov/pub/seg/ |
| CENSOR | ftp://ncbi.nlm.nih.gov/repository/repbase/censor/ |
| XBLAST | http://blast.wustl.edu/pub/xblast/ |
| **Weight matrix, HMM definition** | |
| Gibbs sampler | ftp://ncbi.nlm.nih.gov/pub/neuwald/gibbs11_93/ |
| WCONSENSUS | ftp://beagle.colorado.edu/pub/Consensus/ |
| SAM(buildmodel) | http://www.cse.ucsc.edu/research/compbio/sam.html |
| **Weight matrix, HMM search** | |
| MATRIX SEARCH | ftp://beagle.colorado.edu/pub/imd-1.1.tar.gz |
| MatInspector | ftp://ariane.gsf.de/pub/unix/matinspector/ |
| HMMER(hmmpfam,hmmsearch) | http://hmmer.wustl.edu/ |
| **Promoter prediction** | |
| PROMOTER SCAN | ftp://biosci.umn.edu/pub/proscan/ |
| PromFD | ftp://beagle.colorado/edu/pub/PromFD.tar |
| **RNA motif search** | |
| RNABOB | http://www.genetics.wustl.edu/eddy/software/#rnabob |
| tRNAscan-SE | http://www.genetics.wustl.edu/eddy/tRNAscan-SE/ |
| Palingol | http://www.abi.snv.jussieu.fr/cgi-bin/wrap/viari/Palingol/ |
| **DNA bending and curvature** | |
| BEND and BEND_TRI | http://www.scripps.edu/pub/goodsell/research/bend/ |
| CURVATURE | ftp://sgjsl.weizmann.ac.il/pub/Curvature/ |
| **Feature visualization** | |
| Lalnview | ftp://expasy.hcuge.ch/pub/lalnview/ |
| SEView | ftp://ftp.isrec.isb-sib.ch/sib-isrec/SEView/ |

Tandem Repeat Finder (Benson 1997), both available through a WWW interface. Low complexity regions, for example oligopurine:oligopyrimidine tracts, are detected by programs such as SEG (Wootton and Federhen, 1993). Note that the default parameters of SEG are optimized for protein sequences. For DNA analysis, specification of window size 20 and complexity threshold 0.8 is more appropriate.

**Table 8.2**    Specialized databases for gene feature identification.

| Database | URL | References |
|---|---|---|
| Repbase | http://charon.girinst.org/~server/repbase.html | Jurka (1998) |
| Simple repeats | ftp://ncbi.nlm.nih.gov/repository/repbase/REF1/simple.ref | Jurka & Pethiyagoda (1995) |
| Small RNA Database | http://mbcr.bcm.tmc.edu/smallRNA/smallrna.html | Gu *et al.* (1998) |
| HOVERGEN | http://pbil.univ-lyon1.fr/databases/hovergen.html" | Duret *et al.* (1994) |
| EPD | http://www.epd.unil.ch | Cavin *et al.* (1998) |
| TRANSFAC | http://transfac.gbf.de/TRANSFAC/ | Heinemeyer *et al.* (1998) |
| TRRD | http://www.bionet.nsc.ru/trrd/ | Kel' *et al.* (1997) |
| OOTFD | http://www.isbi.net/ | Ghosh (1998) |
| RegulonDB | ftp://www.cifn.unam.mx/Computational_Biology/regulondb/ | Huerta *et al.* (1998) |
| IMD | ftp://beagle.colorado.edu/pub/imd_1.1.tar.gz | Chen *et al.* (1995) |
| DNA structure library | ftp://transfac.gbf.de/pub/structure_library/ | Lavery & Sklenar (1988) |
| NUCLEOSOME | ftp://ftp.ebi.ac.uk/pub/databases/nucleosomal_dna/ | Ioshikhes & Trifonov (1993) |

Often, the purpose of identifying repeats and low complexity segments in genomic sequences is masking such regions so that they will not interfere with subsequent analyses. The BLAST program suite includes the filtering utility Dust which automatically masks compositionally biased subsequences and simple repeats in the query sequence. For mammalian gene sequences, this treatment is usually not sufficient because of the high abundancy of SINE and LINE elements in these genomes. For example, a human sequence containing an Alu element will match over 20 000 other Alu elements in the current nucleotide sequence database. The automatic processing of these matches will not only take a lot of computing time but also produce a very large and, for all practical purposes, unmanageable output list. In order to detect the more subtle and biologically interesting similarities, it is therefore essential to mask these elements prior to a sequence similarity search against a comprehensive sequence database. Among the tools which can be used for this purpose, is

**Table 8.3** Selected WWW servers for gene feature identification.

| Service | URL | Reference (Author) |
|---|---|---|
| **Repeat** | | |
| Tandem | | Benson (1997) |
| Repeats Finder | http://c3.biomath.mssm.edu/trf.upload.form.html | Sagot & Myers (1998) |
| Satellites | http://bioweb.pasteur.fr/seqanal/interfaces/satellites.html | |
| RepeatMasker | http://ftp.genome.washington.edu/cgi-bin/RepeatMasker | (A.F.A. Smit & P. Green) |
| **Motifs, control elements** | | |
| PatSearch (TransFac) | http://transfac.gbf.de/cgi-bin/patSearch/patsearch.pl | Heinemeyer *et al.* (1998) |
| TESS | http://www.gsf.de/cgi-bin/mastersearch.pl | Schug & Overton (1997) |
| MatInspector (TransFac) | http://agave.humgen.upenn.edu/utess/tess31 | |
| Signal Scan (IMD) | http://bimas.dcrt.nih.gov:/molbio/matrixs/ | Quandt *et al.* (1995) Prestridge (1996) |
| **Promoters, composite elements** | | |
| FunSiteP | http://transfac.gbf.de/dbsearch/funsitep/fsp.html | Kondrakhin *et al.* (1995) |
| PROMOTER SCAN II | http://biosci.cbs.umn.edu/software/proscan/promoterscan.htm | Prestridge (1995) |
| Pol3scan | http://irisbioc.bio.unipr.it/pol3scan.html | Pavesi *et al.* (1994) |
| FastM | http://www.gsf.de/cgi-bin/fastm.pl | Heinemeyer *et al.* (1998) |
| **DNA, RNA structure:** | | |
| bend.it (for DNA) | http://www.2.icgeb.trieste.it/~dna/bend_it.html | Munteanu *et al.* (1998) |
| tRNAscan-SE | http://www.genetics.wustl.edu/eddy/tRNAscan-SE/ | Lowe & Eddy (1997) |
| **Integrated tools** | | |
| Grail | http://avalon.epm.ornl.gov/Grail-bin/EmptyGrailForm | Uberbacher *et al.* (1996) |
| TRADAT | http://www.itba.mi.cnr.it/tradat/ | (L. Milanesi) |
| NIX | http://menu.hgmp.mrc.ac.uk/Nix/ | (M.J. Bishop) |

CENSOR, a program by Jurka *et al.* (1996) designed to be used together with the Repbase. A similar effect can be achieved by using BLAST as the search engine, and subsequently processing the output from a Repbase search with the program XBLAST (Claverie and States 1993). The RepeatMasker tool, which is available on a WWW server, uses a more sensitive similarity search program and thus is presumably more effective in detecting matches to ancient, more divergent repeat families, such as the MIR family.

For pairwise sequence comparisons, the programs Dotter (Sonnhammer & Durbin, 1995) and Lalign (FASTA package) can be recommended. These tools are very useful for identifying conserved regions between pairs of homologous noncoding sequences such as orthologous promoter regions from different species. Dotter is a powerful dot matrix program allowing for interactive adjustment of grey-scale parameters, visualization of matching sequence segments, and dynamic change of the window size (see Fig. 8.1). Lalign generates local sequence alignments using the Smith–Waterman algorithm. An important advantage over other local alignment programs is its capability to find multiple local matches between two sequences in one run. The alignment scoring system is provided by a parameter file which can easily be changed manually using a text editor. PRSS, another programme of the FASTA package, empirically determines the significance threshold for alignment scores reported by Lalign by the shuffling method. HOVERGEN (Duret *et al.*, 1994) is an important database for comparative genomics, providing multiple alignments of highly conserved noncoding sequence regions of homologous vertebrate genes.

There are numerous tools for identifying gene control elements and TF-binding sites in DNA sequences. TRANSFAC is the most important database in this field, providing information on specific DNA control elements, as well as consensus sequence and matrix descriptions of the sequence motifs recognized by the cognate gene regulatory proteins. Most WWW-based tools for searching gene control elements use this information resource in one or the other way. A weight matrix library derived from TRANSFAC can be searched with the program MatInspector (Quandt *et al.*, 1995) also accessible as a WWW service. IMD is an alternative matrix collection, calculated in a different way but also partly derived from TRANSFAC, which is accessed by the program MATRIX SEARCH (Chen *et al.*, 1995) and the SIGNAL SCAN server (Prestridge 1996). Additional WWW servers for searching gene regulatory elements are listed in Table 8.3. OOTFD (Ghosh 1998) is an object-oriented database providing information similar to TRANSFAC. Data on prokaryotic gene regulatory elements can be found in RegulonDB (Huerta *et al.*, 1998).

Recently, a number of promoter prediction programs and WWW servers have become available. Examples are FunSiteP (Kondrakhin *et al.*, 1995), PROMOTER SCAN II (Prestridge 1995), and PromFD (Chen *et al.*, 1997). These tools are not yet reliable on their own but may provide useful indications in conjunction with other information, for example a CpG island prediction made by Grail (Uberbacher *et al.*, 1996). The FastM server (Heinemeyer *et al.*, 1998) allows the user to scan a DNA sequence for two regulatory elements occurring within a short interval. This represents a very promising approach to reduce the notoriously high false-positive rates of TF-binding site search tools.

Experimentalists studying the binding-specificity of a newly discovered transcription factor may be interested in tools to derive a weight matrix from a set of unaligned binding site sequences. The Gibbs sampler (Lawrence *et al.*, 1993) and WCONSENSUS (Hertz & Stormo, 1994) programs can be applied to this end. A useful review of these and related tools has recently been published by Frech *et al.* (1997). More advanced users may prefer an HMM training program such as buildmodel from the SAM package. HMM-based algorithms are particularly useful for deriving composite binding-site descriptions consisting of two or more weight matrix blocks separated by spacers of variable length. Note, however, that effective usage of a tool like buildmodel requires a thorough understanding of the function of the parameters which control the iterative sequence realignment process.

There are specialized programs for searching certain classes of RNA genes. tRNAscan-SE (Lowe & Eddy, 1997) finds tRNA genes in genomic sequences. The RNA POL III promoter prediction Pol3scan server (Pavesi *et al.*, 1994) may also be useful for this purpose because many RNA genes contain internal POL III promoters. The highly conserved ribosomal RNA genes are readily detected by BLAST. However, for small RNAs, a Smith–Waterman search against the small RNA database (Gu *et al.*, 1998) may be more effective. To detect RNA regulatory signals and RNA species where the diagnostic features mainly consist of secondary structure motifs, more specialized programs, such as RNABOB or Palingol (Billoud et al., 1996) are required. RNABOB is faster than Palingol but less versatile because it uses a simpler pattern definition language.

DNA bendability and curvature can be predicted from sequence by using the WWW-server bend.it (Munteanu *et al.*, 1998) or the BEND program (Goodsell & Dickerson, 1994). Both tools rely on parameter sets from the DNA structure library compiled by Lavery & Sklenar (1988). The reliability and biological significance of such predictions remains however controversial. A tool to predict nucleosomes, the preferred binding sites of histone octamers on DNA, is included in the program CURVATURE (Shpigelman *et al.*, 1993). The NUCLEOSOME database (Ioshikhes & Trifonov, 1993), a collection of experimentally mapped nucleosomes, is an important resource for developers of such methods.

Graphic visualization is often helpful and sometimes essential for the interpretation of results produced by sequence analysis programs. Most WWW servers for gene feature identification have their own graphic interfaces. SEView (Junier & Bucher, 1998) is a versatile java applet for interactive display of gene features, which uses a simple sequence element description language and thus could easily be added to existing WWW applications still producing text output. Sequence alignments generated on local computers can be visualized by Lalnview (Duret *et al.*, 1996), which serves also as a Netscape helper application for displaying output generated by a remote server.

For the biologist interpreting gene feature predictions, it would be very convenient to look at the results from different types of analyses simultaneously through the same graphic interface. To achieve this goal, an integration effort is required to make programs from different packages using different formats compatible with each other. The Grail, TRADAT, and NIX servers, each offering a variety of different prediction tools, illustrate the current state of such integration projects.

## 8.5   AN EXAMPLE

This Section illustrates some of the previously introduced tools by an application on real biological data. Two homologous database sequences (accession numbers Z70243 and M16398) containing the 5' flanking regions of the human and mouse interleukin-2 receptor alpha (IL-2R alpha) genes were selected as an example. Both sequences have been the subject of intensive experimental investigations leading to the characterization of several important transcriptional control elements (Lecine *et al.*, 1996). The human sequence contains 8953 bp upstream from the initiation site, the mouse sequence 2545 bp.

Transcription of the IL-2R alpha gene is induced by cytokines and antigens. In cell cultures, two different types of responses to external stimuli have been observed: Antigen in the absence of IL-2 triggers a transient expression of the gene whereas IL-2 administered simultaneously or shortly after antigen, induces high levels of expression over a prolonged period. IL-2 thus appears to be involved in a positive feedback loop regulating its receptor. Experiments have linked transient induction to control elements in the proximal promoter region. The response to IL-2 was shown to be mediated by a short (less than 100 bp long) enhancer segment located at about −3750 and −1350 in the human and mouse IL-2R alpha genes, respectively.

Because repetitive elements can interfere with other types of sequence analysis, the RepeatMasker script was applied first. Several SINEs and LINEs were found (see Fig. 8.3) covering in total 30% of the human and 15% of the mouse sequence, respectively. In addition, one microsatellite $(CAAAA)n$ and three AT-rich regions were identified. The overall G+C content of the human sequence was reported to be 46%, which is close to the genome average and attributes this gene to the H1 isochore class. A search of the database of all public nucleotide sequences (except ESTs) with the masked version of the human sequence revealed homologous 5' flanking regions from pig, sheep, cow, and mouse. A few mRNA sequences matching the short transcribed part at the 3' end of the query were also picked up. It is interesting to compare these results with the corresponding results obtained with the unmasked sequence. With the unmasked version, the search took much longer (about 50 min) and no homologous sequences from other mammalian species were found because

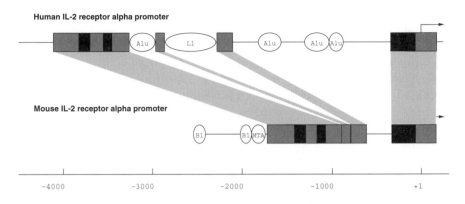

**Figure 8.3**  Gene features identified by similarity search methods. Shown are the interspersed repetitive elements found by RepeatMasker (oval boxes) and conserved sequence regions identified by pairwise sequence alignment (rectangular boxes). The grey areas correspond to the regions found with the mismatch-tolerant default scoring system of the program Lalign. The black areas represent highly conserved regions (>80% identity) selected by the more stringent scoring system described in the text.

the 500 highest-scoring matches contained in the hit list were all matches to human repetitive elements. For human noncoding genomic sequences, masking of SINEs and LINEs prior to BLAST searches is therefore essential but not automatically performed by most public BLAST servers.

The BLAST results clearly indicated that there are conserved regions between the human and mouse 5′ flanking sequences of the IL-2R alpha genes. To localize these regions, a dot matrix comparison was performed using the programme Dotter. As can be seen in Fig. 8.1 (page 146), there are two major conserved areas, one including the experimentally identified IL-2 responsive enhancer and the other one corresponding to the proximal promoter region. Two minor short diagonals can be noted between the two major diagonal lines. Whether they represent homologous sequences is, however, not obvious from the dotplot because their intensities do not exceed the background noise. In order to analyse the conserved regions in more detail and to assess the statistical significance of the dubious signals observed in the dotplot, the sequences were compared with the local alignment program Lalign from the FASTA package. Two different scoring systems were used: the default scoring system (match = 5, mismatch = −4), and a more stringent scoring system (match = 2, mismatch = −4) ensuring a minimal sequence identity of 67%. Scores for 1000 shuffled sequence pairs were generated with the program PRSS in order to determine appropriate significance thresholds for both scoring systems. Surprisingly, the threshold values determined turned out to be quite different: For the default scoring system, an alignment score of 157 was required for significance level $P = 0.01$, whereas for the stringent scoring system, a score as lows as 48 was found to be sufficient. Note that these values depend on the length and the base composition of the sequences compared, and thus have to be computed for each new sequence pair. The weak diagonals observed in the dot matrix proved to be the only significant matches besides the major conserved areas. The stringed scoring system identified a total of three highly conserved sequence segments, each exceeding 80% identity, within the homologous sequence regions delineated with the more mismatch-tolerant method (Fig. 2). The first co-localizes approximately with the experimentally defined enhancer region, the third contains most of the signals that have been implicated in the primary response to antigen stimulation. The function of the middle one remains to be elucidated by experiments.

Three promoter prediction programs, FunSiteP, Promoter Scan, and PromFD, were tested in this case study. The results obtained were rather disappointing. None of the methods was able to locate the known transcription initiation sites correctly. Promoter Scan did not report any promoter, PromFD made one prediction in the human sequence, and FunSiteP made three predictions, two for the human and one for the mouse sequence. The biological significance of these predicted sites has to be considered as unknown; some could correspond to minor upstream transcription initiation sites not yet discovered. In a parallel analysis, the Grail server did not find a CpG island near the major start sites. The conclusion from these results is that it would not have been possible to guess the 5′ end of the two IL-2R alpha genes using sequence feature prediction methods only.

To test the usefulness of transcription binding site prediction methods, the highly conserved sequence segments corresponding to the known enhancer were analysed with three different tools: TESS, Matrix Search, and MatInspector. The results are summarized in Table 8.4. In order to appraise these results, it is necessary briefly to review the known biological facts pertaining to these sequences. In both species, the enhancer region contains three distinct protein-binding sites, as defined by footprinting experiments and electrophoretic mobility shift assays. Site II interacts with a STAT protein (most likely STAT5), and site III is recognized by an Ets family member (Elf-1 in the mouse). Site I was found to have different physiological properties in the two species. In human, this site reportedly binds both STAT and Ets protein family members whereas in mouse it binds only to STAT proteins.

The results obtained with the prediction tools are only partly in agreement with the experimental findings. None of the programs predicted more than four out of the seven known DNA–protein interactions correctly. Moreover, for a sequence of only about 120 bp, the numbers of hits returned by Matrix Search and MatInspector are rather high, suggesting that scanning a longer DNA sequence would produce a very large hit list that would be prohibitively

**Table 8.4**   Prediction of transcription factor binding sites in the IL-2R alpha enhancer regions.

| | Experimental | TESS | Matrix search | MatInspector |
|---|---|---|---|---|
| Human IL-2R alpha enhancer region | | | | |
| Total number of matches | ? | 3 | 20 | 40 |
| True STAT binding sites found | 2 | 1 | 0 | 1 |
| True Ets binding sites found | 2 | 0 | 1 | 1 |
| Mouse IL-2R alpha enhancer region | | | | |
| Total number of matches | ? | 5 | 19 | 31 |
| True STAT binding sites found | 2 | 0 | 0 | 1 |
| True Ets binding sites found | 1 | 1 | 0 | 1 |

time-consuming to analyse. The interpretation of the output listings was further complicated by the fact that the sites reported were not linked in an obvious way to transcription factor families with similar binding specificity. For example, the correctly located Ets-binding sites appeared under names like PU.1, E74A, and EA3. Matrix Search was the only tool which specifically identified site III as an Elk-1 binding site.

Finally, the two sequences were subjected to structural analysis by two different tools. Curvature and bendability predictions were obtained from the Bend.it server. The program CURVATURE was also used for calculating curvature, and in addition to locate nucleosomes. The curvature predictions from the two different tools are difficult to interpret because they are partly in conflict with each other. Nevertheless, both tools agree in suggesting that the sequence regions immediately downstream of the transcription initiation site are relatively straight. The CURVATURE program further predicts strong nucleosomes (histone octamer binding sites) at about position −100 in both promoters, tentatively placing the transcription start site into a linker region. Somewhat surprisingly, the Bend.it server reported low bendability at about the same position which would make DNA wrapping around the histone octamer energetically unfavourable. A better general understanding of the structure–function relationship of promoters and of the role of chromatin in transcription regulation would be required to make sense out of these data.

In summary, the application of a variety of feature prediction tools to biologically well characterized sequences gives some indication of the potential usefulness and current limitations of such methods. In this case, phylogenetic footprinting, i.e. the identification of highly conserved regions between homologous sequences, was very successful in locating important regulatory regions previously found by experiments. The search for interspersed repetitive elements provides an essential road map for further biological characterization. The results obtained using other tools are less conclusive.

# REFERENCES

Altschul S.F., Gish W., Miller W., Myers E.W. & Lipman D.J. (1990) *J. Mol. Biol.* **215**, 403–410.

Altschul S.F., Madden T.L., Schaffer A.A., Zhang J., Zhang Z., Miller W. & Lipman, D.J. (1997) *Nucl. Acids Res.* **25**, 3389–3402.

Antequera F. & Bird A. (1993) *Proc. Natl. Acad. Sci. USA* **90**, 11995–11999.

Benson G. (1997) *J. Comp. Biol.* **4**, 351–367.

Berg O.G. & von Hippel P.H. (1987) *J. Mol. Biol.* **193**, 723–750.

Berg O.G. & von Hippel P.H. (1988) *Trends Biochem. Sci.* **13**, 207–211.

Bernardi G. (1995) *Annu. Rev. Genet.* **29**, 445–476.

Billoud B., Kontic M. & Viari A. (1996) *Nucl. Acids Res.* **24**, 1395–1403.

Brendel V. & Trifonov E.N. (1984) *Nucl. Acids Res.* **12**, 4411–4427.

Bucher P. & Bairoch A. (1994) *ISMB*, **2**, 53–61.

Cavin Perier R., Junier T. & Bucher P. (1998) *Nucl. Acids Res.* **26**, 353–357.

Chen Q.K., Hertz G.Z. & Stormo G.D. (1995) *Comput. Appl. Biosci.* **11**, 563–566.

Chen Q.K., Hertz G.Z. & Stormo G.D. (1997) *Comput. Appl. Biosci.* **13**, 29–35.

Claverie J.M. & Audic S. (1996) *Comput. Appl. Biosci.* **12**, 431–439.

Claverie J.M. & States, D.J. (1993) *Computers Chem.* **17**, 191–201.

Duret L., Dorkeld F. & Gautier C. (1993) *Nucl. Acids Res.* **21**, 2315–2322.

Duret L., Mouchiroud D. & Gouy M. (1994) *Nucl. Acids Res.* **22**, 2360–2365.

Duret L., Gasteiger E. & Perriere G. (1996) *Comput. Appl. Biosci.* **12**, 507–510.

Eddy S.R. (1996) *Curr. Opin. Struct. Biol.* **6**, 361–365.

Fickett J.W. & Hatzigeorgiou A.G (1997) *Genome Res.* **7**, 861–878.

Frech K., Quandt K. & Werner T. (1997) *Comput. Appl. Biosci.* **13**, 89–97.

Gardiner-Garden M. & Frommer M. (1987) *J. Mol. Biol.* **196**, 261–282.

Ghosh D. (1998) *Nucl. Acids Res.* **26**, 360–362.

Goodsell D.S. & Dickerson R.E. (1994) *Nucl. Acids Res.* **22**, 5497–5503.

Grate L. (1995) *ISMB* **3**, 136–144.

Gribskov M., McLachlan A.D. & Eisenberg D. (1987) *Proc. Natl. Acad. Sci. USA* **84**, 4355–4358.

Gu J., Chen Y. & Reddy R. (1998) *Nucl. Acids Res.* **26**, 160–162.

Guigó R. & Fickett J.W. (1995) *J. Mol. Biol.* **253**, 51–60.

Heinemeyer T., Wingender E., Reuter I., Hermjakob H., Kel' A.E., Kel O.V., Ignatieva E.V., Ananko E.A., Podkolodnaya O.A., Kolpakov F.A., Podkolodny N.L. & Kolchanov N.A. (1998) *Nucl. Acids Res.* **26**, 362–367.

Henderson B.R. (1996) *Bioessays.* **18**, 739–746.

Heringa J. (1998) *Curr. Opin. Struct. Biol.* **8**, 338–345.

Hertz G.Z. & Stormo G.D. (1994). In *Bioinformatics and Genome Research: Proceedings of the 3rd International Conference*, Lim, H.A. & Cantor, C.R. (eds). World Scientific Publishing, Singapore, pp. 199–214.

Hertz G.Z. & Stormo G.D. (1996) *Meth. Enzymol.* **273**, 30–42.

Hinton G.E. (1992) *Sci. Am.* **267**(3), 144–151.

Huerta A.M., Salgado H., Thieffry D. & Collado-Vides J. (1998) *Nucl. Acids Res.* **26**, 55–59.

Ioshikhes I. & Trifonov E.N. (1993) *Nucl. Acids Res.* **21**, 4857–4859.

Ioshikhes I., Bolshoy A., Derenshteyn K., Borodovsky M. & Trifonov E.N. (1996) *J. Mol. Biol.* **262**, 129–139.

Junier T. & Bucher P. (1998) *In Silico Biol.* 1, 13–20.

Jurka J. (1998) *Curr. Opin. Struct. Biol.* 8, 333–337.

Jurka J & Pethiyagoda C. (1995) *J. Mol. Evol.* 40, 120–126.

Jurka J., Klonowski P., Dagman V. & Pelton P. (1996) *Comput. Chem.* 20, 119–121.

Kel' A.E., Kolchanov N.A., Kel' O.V., Romashchenko A.G., Anan'ko E.A., Ignat'eva E.V., Merkulova T.I., Podkolodnaia O.A., Stepanenko I.L., Kochetov A.V., Kolpakov F.A., Podkolodnyi N.L. & Naumochkin A.A., (1997) *Mol. Biol.* 31, 521–530.

Kondrakhin Y.V., Kel' A.E., Kolchanov N.A., Romashchenko A.G. & Milanesi L. (1995) *Comput. Appl. Biosci.* 11, 477–488.

Kraus R.J., Murray E.E., Wiley S.R., Zink N.M., Loritz K., Gelembiuk G.W. & Mertz J.E. (1996) *Nucl. Acids Res.* 24, 1531–1539.

Krogh A., Brown M., Mian I.S., Sjolander K. and Haussler D. (1994). *J. Mol. Biol.* 235, 1501–1531.

Lavery R. & Sklenar H. (1988) *J. Biomol. Struct. Dyn.* 6, 63–91.

Lawrence C.E., Altschul S.F., Boguski M.S., Liu J.S., Neuwald A.F. & Wootton J.C. (1993) *Science* 262, 208–214.

Lecine P., Algarte M., Rameil P., Beadling C., Bucher P., Nabholz M. & Imbert J. (1996) *Mol. Cell. Biol.* 16, 6829–6840.

Lipman D.J. (1997) *Nucl. Acids Res.* 25, 3580–3583 (Abstr.).

Lowe T.M. & Eddy S.R. (1997) *Nucl. Acids Res.* 25, 955–964.

Matis S., Xu Y., Shah M., Guan X., Einstein JR., Mural R. & Uberbacher E. (1996) *Comput. Chem.* 20, 135–140.

Mulligan M.E., Hawley D.K., Entriken R. & McClure W.R (1984) *Nucl. Acids Res.* 12, 789–800.

Munteanu M.G., Vlahovicek K., Parthasaraty S., Simon I. & Pongor S. (1998) *Trends Biochem. Sci.* 23, 341–346.

Pavesi A., Conterio F., Bolchi A., Dieci G. & Ottonello S. (1994) *Nucl. Acids Res.* 22, 1247–1256.

Pearson W.R. (1990) *Meth. Enzymol.* 183, 163–198.

Pearson W.R. (1994) *Meth. Mol. Biol.* 24, 307–331.

Pietrokovski S. & Trifonov E.N. (1992) *Gene.* 122, 129–137.

Pollock R. & Treisman R. (1990) *Nucl. Acids Res.* 18, 6197–6204.

Prestridge D.S. (1995) *J. Mol. Biol.* 249, 923–932.

Prestridge D.S. (1996) *Comput. Appl. Biosci.* 12, 157–160.

Quandt K., Frech K., Karas H., Wingender E. & Werner T. (1995) *Nucl. Acids Res.* 23, 4878–4884.

Ryden T.A. & Beemon K. (1989) *Mol. Cell. Biol.* 9, 1155–1164.

Sagot M.F. & Myers E.W. (1998) *J. Comp. Biol.* 5, 539–554.

Schug, J. and Overton, G.C. (1997). *TESS: Transcription Element Search Software on the WWW. Tech. Rep. CBIL-TR-1997-1001-v0.0.* Computational Biology and Informatics Laboratory, School of Medicine, University of Pennsylvania.

Shpigelman E.S., Trifonov E.N. & Bolshoy A. (1993) *Comput. Appl. Biosci.* 9, 435–440.

Smit A.F. (1996) *Curr. Opin. Genet. Dev.* 6, 743–748.

Smith T.F. & Waterman M.S. (1981) *J. Mol. Biol.* 147, 195–197.

Sonnhammer E.L. & Durbin R. (1995) *Gene* 167: GC1–10.

Staden R. (1984) *Nucl. Acids Res.* 12, 505–519.

Stormo G.D., Schneider T.D., Gold L. & Ehrenfeucht A. (1982) *Nucl. Acids Res.* **10,** 2997–3011.

Stormo G.D., Strobl S., Yoshioka M. & Lee J.S. (1993) *J. Mol. Biol.* **229,** 821–826.

Uberbacher E.C., Xu Y. & Mural R.J. (1996) *Meth. Enzymol.* **266,** 259–281.

Trifonov E.N. (1991) *Trends Biochem. Sci.* **16,** 467–470.

Wootton J.C. & Federhen S. (1993) *Comput. Chem.* **17,** 149–163.

Zhang M. & Marr T. (1993) *Comput. Appl. Biosci.* **9,** 499–509.

# 9 Multiple Sequence Alignment

*Des Higgins*

Department of Biochemistry, University College, Cork, Ireland

## 9.1 INTRODUCTION

An example of a multiple sequence alignment is shown in Fig. 9.1. This is a small selection of globin sequences with known tertiary structures. The primary sequences have been lined up to try to maximize the correspondence between the residues at each position. These sequences all have the familiar globin fold that was discovered by John Kendrew and Max Perutz in the 1950s. There are two conserved histidine residues that bind the prosthetic haem group and a conserved core of seven α helices that all of these proteins share. During the course of evolution, residues have been substituted by others (often conserving the biochemical properties of the amino acid side-chains) causing the sequences to diverge from each other, as measured by per cent identity. In this case, the most distant pair is the whale myoglobin (third from the bottom) and the lupin leghaemoglobin (the last sequence). These are only about 10% identical and yet are still recognizably similar when all the sequences are aligned.

There have also been insertions and deletions in various sequences. The evidence for this is the gaps that need to be inserted at some positions in order to maintain the correspondence between the homologous helices. It is usually difficult to tell whether a gap is the result of an insertion or a deletion in a given sequence. There have presumably been more insertion/deletion events in the past that are masked by the current arrangement. If two or more events happen at the same place, it can be difficult to tell exactly what has happened.

So, how does one go about creating an alignment such as this? Ideally, one would line up the seven examples so as to align the corresponding structural parts of the different proteins, if one knew the tertiary structures. Even when some or all of the sequences have known structures, it is an ongoing area of research as to how this is best done. The corresponding situation with align-

*Genetics Databases*
ISBN 0-12-101625-0

CLUSTAL X (1.64b) multiple sequence alignment

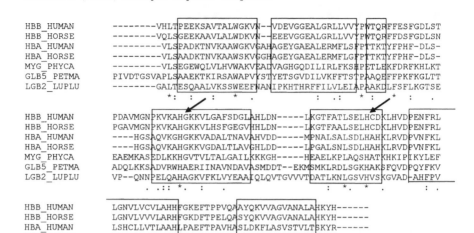

**Figure 9.1**   A multiple alignment of seven globin sequences whose three-dimensional structures are known. The sequence names are from Swiss-Prot and are for two β-globins (HBB_), two α-globins (HBA_), a whale myoglobin (MYG_PHYCA), a cyanohaemoglobin from a lamprey (GLB5_PETMA) and a leghaemoglobin from the lupin (LGB2_LUPLU). The two important histidine residues are marked with arrows and the approximate positions of α-helices are marked with boxes. The symbols under each block of alignment indicate residue conservation (asterisk for identity, colon for strongly conserved and full stop for weakly conserved).

ments of RNA sequences is also complicated where one is trying to align corresponding stems and loops. More commonly, none of the structures are known and one must use alternative methods that will, hopefully, produce alignments that reflect the underlying structural and evolutionary logic. There are basically three methods available: automatic multiple alignment programs, a dedicated alignment editing program, or a combination.

If it were simply a matter of plugging a few sequences into a program and collecting the perfect answer a few moments later, then life would indeed be simple. It can sometimes be like that: the available software is now sufficiently sophisticated to allow users to generate reliable multiple alignments, provided that the sequences are not too divergent and are genuinely related (both by structure and by evolution). The data set must be free of contamination from serious noise such as an amino acid sequence derived from a DNA sequence read in the incorrect way (e.g. including frame shifts). If the user has such a clean data set then the available automatic software can be used to make very good approximations to ideal multiple alignments.

In practice, such data sets are often infected with noise, fragments, errors or contain sequences that are simply too distantly related to align easily. These data sets may cause problems. This is not because the programs are badly written (although this is sometimes the case) but because the user may be asking an impossibly difficult question. In such cases, users must be vigilant and try to watch for clearly nonsensical alignments and try to find and remove sequences that may be responsible for this. This is most easily done as part of an iterative approach. One carries out a series of test alignments and examines both the underlying phylogenetic tree showing the relatedness of the sequences to each other and the intermediate alignments.

Finally, if the user has extra information, such as knowledge of particular active site residues, then this information can be used to help spot alignment errors or can be used to manually edit alignments. One can use automatic methods to generate approximate starting alignments and then edit these to reflect more detailed information or simply to generate an alignment that looks better to the user. One can, of course, use an editor from the start but this can be tedious, with more than a few sequences.

## 9.2   SELECTING THE SEQUENCES TO ALIGN

Typically, sets of sequences will be located in a database using a similarity search method such as FASTA (Pearson & Lipman, 1988) or BLAST (Altschul *et al.*, 1997). Alternatively, sets will be found using keywords and a database browsing tool such as SRS (Etzold *et al.*, 1996) or from a combination of the two. In the former case, the user usually starts with one or two sequences that have come from their own sequence data or from a publication. The user will then use these to search a sequence database such as SwissProt (Bairoch & Apweiler, 1996) using similarity to the query sequence to find related sequences. If there are no related sequences, most similarity searches will still report some possible matches. These will be sequences that are not related to the query sequence but that are the most similar in the database, possibly owing to a biased amino acid composition. For example, sequences that are rich in one or two residues such as proline will appear similar to other proline-rich sequences even though they are not genuinely related. The user will have to decide whether or not to use these possible matches by using common sense and a knowledge of the background biology relating to the query.

If it is very difficult to decide if a particular match is real or not, a multiple alignment between the query sequence and a few of the possible matching sequences can help. If we take the example in Fig. 9.1, we can see that there are several residues conserved between all of the globins in the alignment (such as the haem-binding histidines). These residues show up as conserved vertical stripes and cannot be seen when we just take two of the sequences alone. Here, we would have difficulty deciding if the whale myoglobin and leghaemoglobin were related or not based on the two sequences but the similarity is clear when we examine the entire alignment.

If there are genuinely related sequences in the database, these will hopefully score most highly in the similarity search and go to the top of the output list. This is not guaranteed, however, and the user will still be faced with deciding which sequences to include and which to reject. One simple approach is to take the most related sequences first, make a quick, automatic multiple alignment with these and look for conserved patterns of residues (like the histidines in the globin example). Then, more weakly related hits can be screened quickly because they should also have some of the conserved residues that were found in the earlier alignment.

This will leave the user with a set of sequences to align. If these are reasonably similar and of roughly the same length, then these can often be aligned simply and quickly and in one pass using an automatic method. If the data set contains fragments of sequences (short, incomplete domains) then these can cause problems later and should be used carefully. Further, if the proteins contain multiple domains such as fibronectin or SH3 domains, then it may be wise to try to split these. Attempts may be made to align these separately, unless all the examples have the same domains in the same order along the sequences.

If the sequences were selected using keywords rather than sequence similarity then the user must be particularly careful. There are no guarantees that two sequences with the same keyword will be evolutionarily related (homologous). For example, the same enzymatic function may be carried out by totally different proteins in different taxonomic groups. Further, keywords may be incorrectly assigned to particular sequences, either during the annotation of the sequence by the database curators or by the scientists who reported the sequence.

Finally, the user will end up with a set of sequences that will include certain matches along with some possible ones. These are usually extracted from the database and saved in a single file. These may then be aligned using one of the automatic or manual packages. Some of these packages expect to find the sequences in a certain format; others are more flexible and can read several formats. These formats may come pre-supplied by the software or server used to browse the database so it is worth checking. If an editor or word processor is used to manipulate the format in any way it is usually essential to save the file as 'ASCII text' or 'Text only with linebreaks'. Native word processor file formats are normally completely unreadable by sequence analysis packages.

## 9.3 AUTOMATIC SEQUENCE ALIGNMENT

Given just two sequences, there are fast and effective methods that are guaranteed to find the alignment with the best possible score. These require tables of scores for all possible pairs of matched or mismatched amino acids or nucleotides. The PAM (Point Accepted Mutation) weight matrices of Dayhoff *et al.* (1978) or the BLOSUM amino acid weight matrices of Henikoff & Henikoff (1992) may be used as well as penalties for gaps of different lengths. These are generally derived from the dynamic programming method of Needleman & Wunsch (1970). FASTA and BLAST can be considered as fast approximations of dynamic programming adapted for use as database search tools.

Dynamic programming can be generalized to more than two sequences where one finds the multiple alignment with the best score according to the scoring scheme. Here, one would use an amino acid weight matrix to give high scores to conserved columns of residues and lower scores to columns showing a mixture of residue types. Gaps are scored according to their length and the number of sequences that they occur in and the method will find a multiple alignment with the best possible score. Sequence weights can be used to give more or less weight to particular sequences or pairs of sequences in the alignment (e.g. Thompson *et al.*, 1994a). This is mathematically simple yet powerful, but it is generally considered too slow and demanding on computer resources for routine use. To align more than four or five sequences could take years to carry out given conventional computer power. Clever programming and careful examination of the method were used to make this method practicable for about seven or eight short sequences in the MSA program of Lipman *et al.*, (1989). (This was later made even more efficient by Sandeep Gupta, John Kececiogulu and Alejandro Schaeffer with the fastMSA program.) The MSA program is available from the authors but it requires considerable computer power and memory and a fast workstation will normally be required to run it. It is easier to run this program on a dedicated multiple alignment server where the user pastes the sequences to be aligned into a window and the server carries out the hard work and returns an alignment; examples include the BCM search launcher at Baylor College of Medicine (http://dot.imgen.bcm.tmc.edu: 9331/multi-align/multi-align.html) or the Darwin server at the ETH in Switzerland (http://cbrg.inf.ethz.ch/MultAlign.html.) Despite the improvements in efficiency, the method is still not capable of aligning many sequences.

Most automatic multiple alignments are carried out using a short cut, called 'progressive alignment' by Feng & Doolittle (1987). Here, the alignment is built up gradually from a series of small, computationally inexpensive, alignments. An alignment of two sequences can be carried out in less than 1 s for short sequences of, say, 150 residues (the length of a globin). This gives an alignment between two sequences that maximizes the scores between the pairs of aligned residues and minimizes the number and lengths of the gaps. This can be generalized to allow one to align one sequence against a set of pre-aligned sequences (i.e. to increase the size of an alignment by one sequence) or to align two alignments with each other (see Fig. 9.2). One can use this trick to build up a multiple alignment by aligning larger and larger alignments with each other until all sequences are aligned. The order of alignment (i.e. which sequences are aligned next) could be random or according to some simple scheme where the sequences are sorted according to how similar they are to each other (Barton & Sternberg, 1987). In practice, the most effective way of doing this is to use an evolutionary framework and to align the sequences according to the branching order in a phylogenetic tree of the sequences to be aligned (Taylor, 1988; Corpet, 1988; Higgins & Sharp, 1988).

In the case of our globin example we build up the globin alignment by following a tree of the sequences (Fig. 9.3). First the two β-globins are aligned. These sequences are very similar and can be aligned without any gaps. These are then kept fixed for later alignments and are treated as if they were a single sequence except that, at each position, we have a column of aligned residues instead of just one residue. Next, the two α-globins are aligned and these are also very similar and can be aligned with no gaps. Then the α-globins are aligned with the β-globins. This step aligns four sequences together but only requires the same time as a simple alignment of two sequences. There is no guarantee that there is not a better alignment between the sequences that could be found by exhaustively examining alternatives. In this case, however, the alignment is reasonably certain because the earlier alignments were so simple, the sequences being so closely related. This illustrates an important principle of

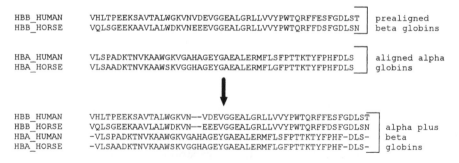

**Figure 9.2**    Illustration of the alignment of two existing alignments using the *N*-terminal 50 or so residues of the human and horse haemoglobin sequences. The β-globins get one short gap of length 2 inserted and the α-globins get three gaps (including the ends) inserted. The pairs are kept fixed during alignment and all four sequences are fixed for the later alignment of sequences.

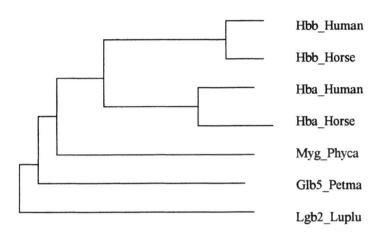

Hbb_Human

Hbb_Horse

Hba_Human

Hba_Horse

Myg_Phyca

Glb5_Petma

Lgb2_Luplu

**Figure 9.3**   The guide tree that is used to make the progressive alignment of the seven globin sequences. Branch lengths are shown roughly to scale. This is a neighbour-joining tree based on separately calculated pairwise alignments of the sequences.

progressive alignment. Accuracy is derived by aligning the most similar sequences first, these being the easiest. Then, when it comes to the more divergent sequences that are harder to align, some information about amino acid conservation and variability at each position is already known from the earlier alignments.

The alignment between the α- and β-globins requires the use of a few gaps. These are inserted into both α- or both β-globins together. One cannot insert a gap in one α-globin without the other. The pre-aligned sequences remain fixed together. After this alignment step we have the four haemoglobins fixed together. These four are then aligned with the myoglobin but this step is still straightforward because myoglobin sequences are relatively similar to haemoglobin sequences. The next step is the alignment of the cyanohaemoglobin with the first five sequences and finally the alignment of the leghaemoglobin with the rest. This latter alignment is difficult because these sequences are very divergent but it is made easier by aligning the last sequence with an existing set of globins rather than just one.

The above procedure was first described in this way by Taylor (1988) but there have been numerous variations and implementations. One of the most widely used packages is the so called Clustal series of programs (Higgins & Sharp, 1988). This has been updated and developed in many ways and is now available as Clustal W (Thompson *et al.*, 1994b) and Clustal X (Thompson *et al.*, 1997). The latter is basically the former with a windows interface and some extra alignment analysis tools. The use of these programs will be described later.

Multiple alignment is an ongoing area of research. Progressive alignment suffers from an inability to correct errors in early alignments as new sequences are added. There has been some work in finding alternative ways of computing the guaranteed global optimal alignment that is obtained from MSA or alignments that approximate this (e.g. Notredame & Higgins, 1996). Recently, a series of methods have been described that rely on finding conserved segments of alignment without gaps. The entire alignment is assembled from these segments (e.g. Morgenstern *et al.*, 1996, WWW site: http://bibiserv.techfak.uni-bielefeld.de/dialign/). The segments may be directly useful in finding local multiple alignments.

## 9.4    USING CLUSTAL W AND CLUSTAL X

Clustal W (Thompson *et al.*, 1994b) is perhaps the most widely used program for carrying out automatic multiple alignments. It is freely available (in source code or executable form) for most computer systems. The easiest way to obtain it is by anonymous FTP from the EMBL/EBI file server (anonymous ftp to ftp.ebi.ac.uk and look in directories of pub/software). It uses a simple text menu system or command-line interface that is reasonably self explanatory and simple to use. It allows one to carry out fully automatic and rapid alignments of dozens of sequences of moderate length using the progressive alignment method. This involves the alignment of all pairs of sequences with each other, the construction of an initial phylogenetic tree using the neighbour-joining method of Saitou & Nei (1987) and finally the alignment of the sequences following the branching order in the tree. The program also features facilities for calculation of trees after multiple alignment and a bootstrap facility for estimating confidence levels on groupings.

There are extensive facilities for controlling the multiple alignments through parameters that guide the placement of gaps in the alignments. These include the usual initial gap opening and gap extension penalties but also include parameters to place the gaps in a residue-specific and secondary structure-specific manner, where this information is available. Pre-aligned sets of sequences can be saved and edited (using alignment editing facilities, not included in Clustal W) and new sequences aligned to these one at a time or in batches. Clustal W is used via a simple text menu system, whereas Clustal X (Thompson *et al.*, 1997) is essentially the same program but it has a much more elegant user interface. Menus may be controlled with a mouse and multicoloured alignments that can dynamically show conservation of amino acid type in columns. Most interestingly, it comes with some simple graphical tools for displaying sections of sequences in the alignment that look peculiar, i.e. where one or more sequences stand out as being very different from the rest, these can be highlighted. This allows one to quickly spot sequences with errors such as result from frame shifts in the original DNA sequence of a protein coding gene. There are also some alignment servers available for use over the Internet (e.g. at the EMBL/EBI http://www.ebi.ac.uk/). It is difficult to set these up with the full functionality of a locally based program but this does offer quick and simple access to the user who only wishes to make occasional alignments.

A full description of the use of Clustal W or Clustal X would require a lengthy document and is far beyond the scope of this brief introduction. Here, I will describe a few of the steps involved in obtaining a simple multiple alignment of the globin sequences we met earlier. I will describe this using the menu system of Clustal W since it is easier to describe but the steps with Clustal X are the same. The first step is the creation of the file with the sequences. The sequences can be in any of the following file formats, provided that the file is simple ASCII text and all of the sequences are in one file only: FASTA format (Pearson & Lipman, 1988), EMBL or SwissProt format, PIR format, GCG/MSF and GCG9/RSF formats, GDE format or Clustal multiple alignment format. Try to save sequences as one of these formats. If this cannot be done, then edit the file and arrange the sequences in FASTA format. This is illustrated in Fig. 9.4 and consists of the seven globin sequences arranged one after another with each sequence delimited by a > symbol in column one. The characters immediately after each > symbol are read as the name for the following sequence.

Clustal W and Clustal X can read in a full multiple alignment. Some of the input file formats are designed for use solely with multiple alignments to begin with (e.g. Clustal and GCG/MSF formats). The other formats can be used to store full alignments by the addition of hyphens to signify gaps. This might seem odd since the program is designed to produce alignments. However, it is useful for three reasons. First, one can read old alignments in order to calculate new phylogenetic trees; trees are normally calculated from full alignments. Second, one can read in an alignment in one format and immediately obtain the same alignment in a new format; there are five different possible alignment formats produced by the program after alignment. Finally, one can read in an old alignment and add new sequences to these using so called 'profile alignment'. As the sequences are read, the program will read any hyphen as a gap (except in GCG/MSF format where full stop is used to signify a gap). Any spaces or numbers are ignored. Any alphabetic characters are read as residues. The program will attempt to guess whether the sequences are amino acid or nucleic acid. This is done by counting the numbers of A, C, G, T, U and N characters. If 85% of the non-gap characters in a sequence are from these six letters (ignoring case), then the sequence is assumed to be nucleotide. You must not try to mix nucleotide and amino acid sequences. There are no facilities for translating nucleotide sequences and the program will become hopelessly confused, as will the user.

```
>HBB_HUMAN
VHLTPEEKSAVTALWGKVNVDEVGGEALGRLLVVYPWTQRFFESFGDLST
PDAVMGNPKVKAHGKKVLGAFSDGLAHLDNLKGTFATLSELHCDKLHVDP
ENFRLLGNVLVCVLAHHFGKEFTPPVQAAYQKVVAGVANALAHKYH
>HBB_HORSE
VQLSGEEKAAVLALWDKVNEEEVGGEALGRLLVVYPWTQRFFDSFGDLSN
PGAVMGNPKVKAHGKKVLHSFGEGVHHLDNLKGTFAALSELHCDKLHVDP
ENFRLLGNVLVVVLARHFGKDFTPELQASYQKVVAGVANALAHKYH
>HBA_HUMAN
VLSPADKTNVKAAWGKVGAHAGEYGAEALERMFLSFPTTKTYFPHFDLSH
GSAQVKGHGKKVADALTNAVAHVDDMPNALSALSDLHAHKLRVDPVNFKL
LSHCLLVTLAAHLPAEFTPAVHASLDKFLASVSTVLTSKYR
>HBA_HORSE
VLSAADKTNVKAAWSKVGGHAGEYGAEALERMFLGFPTTKTYFPHFDLSH
GSAQVKAHGKKVGDALTLAVGHLDDLPGALSNLSDLHAHKLRVDPVNFKL
LSHCLLSTLAVHLPNDFTPAVHASLDKFLSSVSTVLTSKYR
>MYG_PHYCA
VLSEGEWQLVLHVWAKVEADVAGHGQDILIRLFKSHPETLEKFDRFKHLK
TEAEMKASEDLKKHGVTVLTALGAILKKKGHHEAELKPLAQSHATKHKIP
IKYLEFISEAIIHVLHSRHPGDFGADAQGAMNKALELFRKDIAAKYKELG
YQG
>GLB5_PETMA
PIVDTGSVAPLSAAEKTKIRSAWAPVYSTYETSGVDILVKFFTSTPAAQE
FFPKFKGLTTADQLKKSADVRWHAERIINAVNDAVASMDDTEKMSMKLRD
LSGKHAKSFQVDPQYFKVLAAVIADTVAAGDAGFEKLMSMICILLRSAY
>LGB2_LUPLU
GALTESQAALVKSSWEEFNANIPKHTHRFFILVLEIAPAAKDLFSFLKGT
SEVPQNNPELQAHAGKVFKLVYEAAIQLQVTGVVVTDATLKNLGSVHVSK
GVADAHFPVVKEAILKTIKEVVGAKWSEELNSAWTIAYDELAIVIKKEMN
DAA
```

**Figure 9.4**    The seven globin sequences in FASTA format. The amino acids can be in lowercase with spaces mixed in between and the lines can be of varying length. Each sequence is delimited by a > symbol before its name.

## 9.4.1  Simple multiple alignment

When Clustal W is run, the simple menu in Fig. 9.5 is obtained. This allows the user to get some brief on-line help (available from all the menus) and to exit temporarily from the program to run an external command. The user normally chooses menu item 1 (read sequences from disc) and will be prompted for the name of the file containing the sequences. If this is read correctly, the sequence type (amino acid or nucleotide) number and lengths of the sequences will be reported. The user will then usually choose menu item number 2, to carry out multiple alignments. This gives the menu in Figure 9.6. To quickly carry out the multiple alignment, using defaults for all settings, the user should choose

\* \* \* \* \* \* \* \* \* \* \* \* \* \* \* \* \* \* \* \* \* \* \* \* \* \* \* \* \* \* \* \* \* \* \* \* \* \* \* \* \* \* \* \* \* \* \* \* \* \* \* \* \* \* \* \* \* \* \* \* \*
\* \* \* \* \* \* \* \* \* \* \* CLUSTAL W (1.7) Multiple Sequence Alignments  \* \* \* \* \* \* \* \* \*
\* \* \* \* \* \* \* \* \* \* \* \* \* \* \* \* \* \* \* \* \* \* \* \* \* \* \* \* \* \* \* \* \* \* \* \* \* \* \* \* \* \* \* \* \* \* \* \* \* \* \* \* \* \* \* \* \* \* \* \* \*


1. Sequence Input From Disc
2. Multiple Alignments
3. Profile / Structure Alignments
4. Phylogenetic trees

S. Execute a system command
H. HELP
X. EXIT (leave program)


Your choice:

**Figure 9.5**  The main menu of Clustal W. The user types a number followed by the <Return> key to go to one of the sub menus or H for help or X to exit the program.



\* \* \* \* \* \* MULTIPLE ALIGNMENT MENU \* \* \* \* \* \*


1. Do complete multiple alignment now (Slow/Accurate)
2. Produce guide tree file only
3. Do alignment using old guide tree file

4. Toggle Slow/Fast pairwise alignments = SLOW

5. Pairwise alignment parameters
6. Multiple alignment parameters

7. Reset gaps between alignments? = OFF
8. Toggle screen display       = ON
9. Output format options

S. Execute a system command
H. HELP
or press [RETURN] to go back to main menu


Your choice:

**Figure 9.6**  The multiple alignment menu of Clustal W.

menu item 1, which will result in prompts for two file names. Users can just press return to accept these. One is the file that is to store the guide tree allowing the user to use the same file again or to edit it to change the alignment order. The second is to contain the multiple alignment. The alignment is then carried out and the alignment will appear on the screen and will be sent to the output file.

The alignment is carried out by comparing all pairs of sequences and saving the per cent identity between each pair. The gap penalties that control these alignments can be set from menu item 5 ('Pairwise alignment parameters') but it is not normally worth while doing this. The only possible effect will be on the branching order in the guide tree. These alignments are carried out using full dynamic programming. If there are many sequences, or if the sequences are very long, this can be time consuming. The user can choose to use a faster, more approximate alignment method from Wilbur & Lipman (1993), which is many times faster, by choosing menu item 4 (set to use the slow alignments by default). The fast alignments give a very good approximation to the default ones when the sequences are very similar, although they become less reliable with very distantly related pairs of sequences. The per cent identity scores between all the pairs of sequences will flash (hopefully) past on the screen and then the guide tree is calculated using the neighbour-joining method of Saitou & Nei (1987). Finally, the sequences are aligned according to the branching order in this tree. The parameters that control the final multiple alignment are set from menu item 6 ('Multiple alignment parameters'). These do have an effect on the final alignment but are quite involved to explain in detail. The user can set initial gap opening and extension penalties. These are employed in many complicated ways depending on the lengths of the sequences, how similar the sequences are to each other, whether or not there are gaps present in pre-aligned sets of sequences and so on. The user is referred to the original Clustal W paper (Thompson *et al.*, 1994b) and to the on-line help facility.

One very useful setting is the format for the output file that is controlled by menu item 9 (output format options). Here one can choose the format of the output, allowing the user to use the alignment in a different package such as PHYLIP (Felsenstein, 1985) or allowing the user to edit the alignment using a dedicated editor. The submenu is shown in Fig. 9.7. One can choose to turn on or off the output of five different file formats (files can be produced in more than one format at the same time). One extremely useful device is offered by menu item 9 that offers the user the ability to 'create alignment output file(s) now'. If this is chosen, an output alignment file will be produced immediately, corresponding to each of the file formats that are toggled on in the menu. This would not normally be useful if the sequences are not already aligned. However, it is extremely useful to help change the format of an old multiple alignment. One can read it in as usual from the main input menu and go straight to the file format options menu without carrying out a new multiple alignment. Here, the user can then choose to create a new alignment file in one

********* Format of Alignment Output *********

1. Toggle CLUSTAL format output   = ON
2. Toggle NBRF/PIR format output   = OFF
3. Toggle GCG/MSF format output   = OFF
4. Toggle PHYLIP format output   = OFF
5. Toggle GDE format output   = OFF

6. Toggle GDE output case   = LOWER
7. Toggle CLUSTALW sequence numbers = OFF
8. Toggle output order   = ALIGNED

9. Create alignment output file(s) now?

0. Toggle parameter output   = OFF

H. HELP


Enter number (or [RETURN] to exit):

**Figure 9.7**   The multiple alignment output formatting menu of Clustal W.


of the five available formats. One can also choose to put sequence numbers on the output. These give the numbers of the residues at the ends of the lines of alignment. Finally, the default order for the sequences in the alignment will reflect the order that the sequences are aligned in the guide tree. This can be changed to be the same as the order in the input file.

## 9.4.2 Phylogenetic trees

The main menu leads to submenus for calculating phylogenetic trees from multiple alignments and for the alignment of existing alignments (profile alignment). The trees cannot be displayed by Clustal W (or Clustal X) but can be viewed using dedicated tree-viewing programs. Two good examples of the latter are NJplot (written by Manolo Gouy, Lyon, France), which is distributed with Clustal, and Treeview by Rod Page (1996). The latter can be downloaded from the WWW at http://taxonomy.zoology.gla.ac.uk/rod/treeview.html. If users really want to get to grips with making their own trees then this author highly recommends two packages. These are the PHYLIP package (Felsenstein,

1985), which is available for all computers from the WWW address http://evolution.genetics.washington.edu/phylip.html, and Phylo_Win (Galtier *et al.*, 1996), which is available for UNIX machines by anonymous ftp from biom3.univ-lyon1.fr in directory pub/mol_phylogeny. In this case, the user will generate the alignments using Clustal and analyse these to generate the phylogenies using the above dedicated packages that offer a far greater range of techniques. The tree making facilities in Clustal W and Clustal X are, nonetheless, convenient. In order to generate a multiple alignment in the first place, a tree is used. This is stored in a file whose name ends in '.dnd' for dendrogram and is calculated using initial alignment scores between all pairs of sequences. These trees can be viewed and one can try to edit them to manipulate the order of alignment but they are not normally very useful in phylogenetic analysis. They will show the overall relatedness of the sequences but will not necessarily be very accurate. To get more accurate trees, one should derive the tree from a complete multiple alignment (see below). These trees are sent to files whose names end in '.ph' or '.phb' and are in the same format as the dendrogram files above. In simple cases, the trees produced by Clustal W and Clustal X will be very useful and acceptable for publication, provided that the fine details of the phylogeny are not critical. If the purpose of the analysis is the phylogeny and if the details are critical, then one should consider alternative packages.

To generate a tree, the user must already have a multiple alignment ready. This can be in memory directly after the calculation of a multiple alignment. It will stay there until the program is closed, a new data set is read in or the alignment procedure is repeated (perhaps with different settings). Alternatively, the alignment can just be read in as a complete alignment using the sequence input option from the main menu. One chooses item 4 from the main menu to get to the tree menu shown in Fig. 9.8. Here, menu item 1 also allows one to input an alignment. Its usage is identical to the sequence and alignment input items in all the other menus. When the alignment is ready, the user can just use menu item 4 to calculate the tree. Sadly, the tree cannot be drawn as such, it is merely calculated but it can be displayed using other packages such as NJplot or Treeview. The tree is calculated by taking every pair of sequences in the multiple alignment and estimating the mean number of differences per aligned site. Initially, this is a number between zero (no differences for identical sequences) and one (a pair of aligned sequences with no identical residues). These distances can be used to make a tree with the neighbour-joining method but there are two parameters that help control the way the distances are calculated.

One can choose to remove from the analysis all positions in the multiple alignment where any sequence has a gap. This ignores potentially useful information but has the beneficial effect of also removing those alignment positions that are most difficult and problematic, i.e. where the alignment is dubious and impossible to judge. There is one extremely important proviso here and that is that this setting may have unfortunate side-effects if some of the sequences are very short. It is not wise to try to make accurate trees using a mixture of full-

\*\*\*\*\*\* PHYLOGENETIC TREE MENU \*\*\*\*\*\*


1. Input an alignment
2. Exclude positions with gaps?      = OFF
3. Correct for multiple substitutions? = OFF
4. Draw tree now
5. Bootstrap tree
6. Output format options

S. Execute a system command
H. HELP
or press [RETURN] to go back to main menu


Your choice:

**Figure 9.8**   The phylogenetic tree menu of Clustal W.


length sequences and short fragments but if one does this, one should be careful of this setting. The second parameter that can be set is to attempt to correct the initial distances for 'multiple hits'. The protein evolution model of Dayhoff *et al.* (1978) is used to convert an observed distance between two sequences to an estimated number of substitutions that have actually occurred during evolution. An approximation to this correction is given in a simple formula by Kimura (1993) but this only works up to an observed distance of 0.82 observed amino acid differences per site. This correction will take distances and stretch them as an attempt at compensating for the considerable number of hidden changes that we cannot see in current sequences. The effect on small distances is insignificant but becomes greater and greater as the distances grow. At very large distances (e.g. 0.8 or more) the correction may become unreliable. Finally, one can make trees with bootstrap confidence measures for all of the groupings in the tree (Felsenstein, 1985). These help to highlight which groupings in the tree are well supported, given the alignment and the method used to make the tree and which are unreliable.

A tree, produced by Clustal W, for the globin example is shown in Fig. 9.3. The tree is drawn with all branch lengths to scale; the length of each branch is drawn proportional to the estimated 'amount' of evolution that has happened along that branch since the common ancestor. Initially, the tree is unrooted, i.e. we do not know the position of the root or the hypothetical starting point of the sequence divergence. The position of the root can often be guessed at on biological grounds as it is here where the root is placed between the plant leghaemoglobin and the rest. In this case, the plant sequence is said to be an outgroup to the rest.

## 9.4.3    Profile alignment

By profile alignment, I simply mean the alignment of existing alignments, although the original meaning was introduced in the context of a database similarity search (Gribskov *et al.*, 1987). When I refer to a profile here, I simply mean an existing alignment that is in a file (e.g. a Clustal '*.aln' file). An existing alignment can be considered to have just one sequence in the case of a single sequence file. From the main menu of Clustal W one can get to the profile alignment menu shown in Fig. 9.9. This allows the user to align new sequences to an existing alignment. The user must input the name of a file containing the first profile (menu item 1). This is profile_1 and can be a single sequence or an alignment of a number of sequences. This alignment might have been carried out earlier by the user using Clustal W or might have been carefully edited and manipulated. If the user has access to information about the secondary structure of one or all of the sequences in the profile, this is potentially invaluable for directing the placement of gaps. Gaps should be easiest to

****** PROFILE AND STRUCTURE ALIGNMENT MENU ******


1. Input 1st. profile
2. Input 2nd. profile/sequences

3. Align 2nd. profile to 1st. profile
4. Align sequences to 1st. profile (Slow/Accurate)

5. Toggle Slow/Fast pairwise alignments = SLOW

6. Pairwise alignment parameters
7. Multiple alignment parameters

8. Toggle screen display        = ON
9. Output format options
0. Secondary structure options

S. Execute a system command
H. HELP
or press [RETURN] to go back to main menu


Your choice:

**Figure 9.9**   The profile alignment menu of Clustal W.

place between strands or helices and hardest to place in the middles of strands or helices. The secondary structure information can be included in the profile in a special format that is described in the on-line help and the relationship between this and the gap penalties can be set from menu item 0.

The whole point of this menu is to then allow the user to add new sequences to profile_1. This can be done in two ways. First, the user might simply read in a second profile (of one or more aligned sequences) and align it in one step to the first. Second, the user could input a set of unaligned sequences that are to be aligned to the first profile, one at a time. The first method is used to build up a large alignment in controlled stages. The second method is used to quickly add a few new sequences to an existing alignment. Either way, the user must choose menu item 2 to input the sequence(s) from the second profile. When this is done the user can align these to the first profile using menu items 3 and 4. The alignments are carried out using exactly the same parameters and file formats as used in the multiple alignment menu. The facilities for doing this in Clustal X are more sophisticated and use a more attractive interface but the principles are the same.

## 9.5   EDITING AND VIEWING MULTIPLE ALIGNMENTS

There are some extra viewing capabilities that are possible with Clustal X that are absent from Clustal W. The alignments that are produced by Clustal W and X are sent to simple text files. These show the sequences aligned in blocks and use symbols to indicate residues that are exactly conserved in columns or which show some degree of biochemical conservation in the columns, e.g. columns of acidic residues (aspartate and glutamate) or with aromatic side-chains (phenylalanine, tyrosine and tryptophan). Clustal X can show the multiple alignment on the screen and make use of colours to indicate residue conservation. One simple method of indicating conservation is to divide the amino acids into groups according to side-chain properties and allocate a different colour for each. This is crude and simplistic as some amino acids belong to more than one group. Some alignment positions will also clearly show conservation of an amino acid property while others will show a mixture. Clustal X has the ability to dynamically choose whether or not to colour a residue depending on the other residues in the column. This is done dynamically after alignment and is changed if the alignment is altered (e.g. by the later addition of new sequences). Clustal X also has the ability to highlight sections of sequence that are clearly outliers compared with the rest of the data set at those alignment positions. This is useful for picking up sequences that are partly (or completely) unrelated to the rest.

There are more sophisticated facilities available in other packages for marking and annotating alignments. For example, Boxshade by Kay Hoffman and Michael Baron is available free for all computer operating systems and provides extensive facilities for reading alignments in different formats and writing them out again and has many different formats with various types of boxes and shading to indicate residue conservation. Information is available from the WWW at: http://www.isrec.isb-sib.ch/software/BOX_form.html. The AMAS package of Livingstone & Barton (1993) is available on the Internet at http://barton.ebi.ac.uk/servers/amas_server.html and provides extensive online facilities for analysing multiple alignments in order to detect interesting conservation patterns. It does not read Clustal format alignments directly but can read the PIR format files produced by Clustal.

If one deals extensively with multiple alignments, it is extremely useful to use multiple alignment editing software. One can use conventional editors or text editors but these cannot easily deal with wrapping blocks of multiple alignment after inserting or deleting sections of alignment. Dedicated alignment editors can move large gaps quickly and can be persuaded to colour columns of residues depending on the degree of conservation and/or residue type. Some of these will read many formats and are available for several operating systems; some will even interact directly with Clustal. For example, Seaview (Galtier *et al.*, 1996; available by anonymous ftp from biom3.univ-lyon1.fr in directory pub/mol_phylogeny) and SeqPup (Java program written by Don Gilbert and available from http://iubio.bio.indiana.edu/1/IUBio-Software%2bData/mobio/seqpup/) allow the user to quickly edit alignment output from Clustal and get sections realigned.

## REFERENCES

Altschul S.F., Madden T.L., Schaeffer A.A., Zhang J., Zhang A., Miller, W. & Lipman D.J. (1997) *Nucl. Acids Res.* **25**, 3398–3402.
Bairoch A. & Apweiler R. (1996) *Nucl. Acids Res.* **24**, 21–25.
Barton G.J. & Sternberg M.J.E. (1987) *J. Mol. Biol.* **198**, 327–337.
Corpet F. (1988) *Nucl. Acids Res.* **16**, 10881–10890.
Dayhoff M.O., Schwartz R.M. & Orcutt B.C. (1978) In *Atlas of Protein Sequence and Structure.* vol. 5 Suppl. 3, Dayhoff M.O. (ed.). NBRF, Washington, p. 345.
Etzold T., Ulyanov A. & Argos P. (1996) *Meth. Enzymol.* **266**, 114–128.
Felsenstein J. (1985) *Evolution* **39**, 783–791.
Feng D.-F. & Doolittle R.F. (1987) *J. Mol. Evol.* **25**, 351–360.
Galtier N., Gouy M. & Gautier C. (1996) *Comp. Appl. Biosci.* **12**, 543–548.
Gribskov M., McLachlan A.D. & Eisenberg D. (1987) *Proc. Natl. Acad. Sci. USA* **84**, 4355–4358.
Henikoff S. & Henikoff J.G. (1992) *Proc. Natl. Acad. Sci. USA* **89**, 10915–10919.
Higgins D.G. & Sharp P.M. (1988) *Gene* **73**, 237–244.

Kimura M. (1993) *The Neutral Theory of Molecular Evolution.* Cambridge University Press, Cambridge.
Lipman D.J., Altschul S.F. & Kececioglu J.D. (1989) *Proc. Natl. Acad. Sci. USA* **86**, 4412–4415.
Livingstone C.D. & Barton G.J. (1993) *Comput. Appl. Biosci.* **9**, 745–756.
Morgenstern B., Dress A. & Werner T. (1996) *Proc. Natl. Acad. Sci. USA* **93**, 12098–12103.
Needleman S.B. & Wunsch C.D. (1970) *J. Mol. Biol.* **48**, 443–453.
Notredame C. & Higgins D.G. (1996) *Nucl. Acids Res.* **24**, 1515–1524.
Page R. D. M. (1996) *Comp. Appl. Biosci.* **12**, 357–358.
Pearson W. R. & Lipman D.J. (1988) *Proc. Natl. Acad. Sci. USA* **85**, 2444–2448.
Saitou N. & Nei M. (1987) *Mol. Biol. Evol.* **4**, 406–425.
Taylor W.R. (1988) *J. Mol. Evol.* **28**, 161–169.
Thompson J.D., Higgins D.G. & Gibson T.J. (1994a) *Comp. Appl. Biosci.* **10**, 19–29.
Thompson J.D., Higgins D.G. & Gibson T.J. (1994b) *Nucl. Acids Res.* **22**, 4673–4680.
Thompson J.D., Gibson T.J., Plewniak F., Jeanmougin F. & Higgins D.G. (1997) *Nucl. Acids Res.* **25**, 4876–4882.
Wilbur W.J. & Lipman D.J. (1993) *Proc. Natl. Acad. Sci. USA* **80**, 726–730.

# 10 On-line Resources for RNA Science

*Kathleen Triman*

Department of Biology, Franklin & Marshall College, Lancaster, PA, USA

## 10.1 INTRODUCTION

On-line resources for RNA science provide world-wide access to RNA data. This chapter provides an overview of available resources and is divided into three sections:

(1) Specialized RNA-related databases,
(2) Tools for analysis: RNA structure and prediction, and
(3) Future directions.

The first issue of *Nucleic Acids Research* devoted entirely to database announcements was published on 1 July 1993. Seven of the 24 publications in that issue announced RNA-related databases. At least 22 RNA-related databases have been announced and updated in the five database issues published between 1993 and 1998 (Roberts, 1998). These databases (examples described below) define a wide range of on-line resources devoted to specialized RNA data. It is recommended that the reader consult original references cited here for detailed information about individual resources.

## 10.2 SPECIALIZED RNA-RELATED DATABASES

### 10.2.1 5S rRNA Data Bank

One of the first RNA database announcements described the BERLIN database of 5S rRNA sequences (Specht *et al.*, 1991). The updated version of the

*Genetics Databases*
ISBN 0-12-101625-0

compilation of 5S rRNA nucleotide sequences, the 5S rRNA Data Bank, contains 1622 primary structures of 5S rRNA organized according to the taxonomic position of the organisms from which the sequences were obtained (Szymanski *et al.*, 1998). The curators of 5S rRNA Data Bank are Maciej Szymanski, from the Institute of Bioorganic Chemistry at the Polish Academy of Sciences in Poznan, and Volker Erdmann, from the Institute for Biochemistry at the Free University in Berlin, Germany.

## 10.2.2  Compilation of tRNA sequences and sequences of tRNA genes

The compilation of tRNA sequences contains 3279 sequences of tRNAs and tRNA genes (Sprinzl *et al.*, 1998). An alignment, compatible with the tRNA phylogeny and known three-dimensional structures of tRNA, is also provided. The curator of this database is Mathias Sprinzl from the Laboratory for Biochemistry at Bayreuth University in Germany.

## 10.2.3  The Signal Recognition Particle Database (SRPDB)

The Signal Recognition Particle Database contains 99 annotated sequences of signal recognition particle RNAs, grouped phylogenetically and aligned (Larsen *et al.*, 1998). The curator of the SRPDB is Christian Zwieb from the Department of Molecular Biology at the University of Texas Health Science Center in Tyler, TX, USA.

## 10.2.4  The Ribosomal Database Project

The Ribosome Database Project (RDP) offers aligned and phylogenetically ordered rRNA sequences, as well as a variety of software for analysis of the data (Larsen *et al.*, 1993). The curator of the Ribosome Database Project is Niels Larsen from the Department of Microbiology at Michigan State University in East Lansing, MI, USA.

GenCANS (Gene Classification Artificial Neural System), a system originally designed for the automatic classification of protein data, was extended to classify ribosomal RNA sequences from RDP (Wu & Shivakumar, 1994). The curator of the GenCANS-RDP System Web site is Cathy Wu from the Bioinformatics Research Group at the University of Texas Health Center at Tyler, TX, USA.

## 10.2.5    Database on the structure of small sub-unit ribosomal RNA

The Antwerp database on small subunit (SSU) rRNA structure contains 2361 eukaryotic, 5487 bacterial, 238 archaeal, 98 plastid and 445 mitochondrial sequences (van de Peer *et al.*, 1998). The sequences are stored as an alignment. This database is one of two databases of rRNA available on the rRNA World Wide Web (WWW) server. The server also offers a variety of tools that facilitate the analysis of sequences available at the site.

## 10.2.6    Database on the structure of large ribosomal subunit RNA

The Antwerp database on large subunit ribosomal RNA (LSU rRNA) contains a total of 496 sequences, mostly from mitochondria and bacteria (de Rijk *et al.*, 1998). The curator of both databases available from the rRNA WWW server is Peter de Rijk from the Department of Biochemistry at the University of Antwerp in Belgium.

## 10.2.7    Collection of small subunit (16S and 16S-like) ribosomal RNA structures

Secondary and tertiary structure interactions determined with comparative analysis for a collection of 90 phylogenetically diverse 16S and 16S-like rRNAs are available from Robin Gutell, from the Institute for Cellular and Molecular Biology, University of Texas at Austin, USA.

## 10.2.8    Compilation of large subunit (23S and 23S-like) ribosomal RNA structures

Secondary structure diagrams for 80 phylogenetically diverse 23S and 23S-like rRNAs are also available (Gutell *et al.*, 1993).

## 10.2.9    A comparative database of group I intron structures

Secondary structures for Group I introns represent the third major category of structures available from Gutell (Damberger & Gutell, 1994).

## 10.2.10    The translational signal database (TransTerm)

TransTerm contains more than 97 500 coding-sequence initiation and termination contexts compiled from GenBank (Dalphin *et al.*, 1998). The database is being extended to include full sequences of the 5'-untranslated region (UTR), 3'-UTR and coding sequence, in addition to the start- and stop-codon sequence contexts available in previous versions. The curator of TransTerm is Mark Dalphin from the Department of Biochemistry and Centre for Gene Research at the University of Otago in Dunedin, New Zealand.

## 10.2.11    Small RNA Database

The Small RNA Database is a compilation of sequences of small-size RNAs that are not directly involved in protein synthesis (Gu et al., 1998). The published sequences originate from nuclear, nucleolar, cytoplasmic and mitochondrial small RNAs from eukaryotic organisms, as well as small RNAs from prokaryotes and viruses. The 600 small RNAs are grouped as follows: (1) small capped RNAs, (2) non-capped small RNAs, and (3) viral small RNAs. The curator of the Small RNA Database is Jian Gu from the Department of Pharmacology at Baylor College of Medicine in Houston, TX, USA.

## 10.2.12    Ribosomal RNA mutation databases

The ribosomal RNA mutation databases include six separate files of data. These are 16S and 23S rRNA alterations identified in *Escherichia coli* (16SMDB and 23SMDB), 16S-like and 23S-like alterations identified in organisms other than *E. coli* (16S-likeMDB and 23SMDB), 295 alterations combined from 16S and 16S-like rRNA (16SMDBexp), and 271 alterations combined from 23S and 23S-like rRNA 23SMDBexp (Triman *et al.*, 1998). The data include position, nucleotide identity, phenotype and literature citations. This author is the curator of the ribosomal RNA mutation databases (Kathleen Triman, Department of Biology at Franklin and Marshall College in Lancaster, PA, USA).

## 10.2.13    The Ribonuclease P Database

The RNAse P Database is a compilation of RNAse P RNA subunit sequences, alignments, secondary structures, and three-dimensional models (Brown, 1998). The curator of the RNAse P Database is James Brown from the

Department of Microbiology at North Carolina State University in Raleigh, NC, USA.

## 10.2.14   The Viroid and Viroid-like RNA Database

The Viroid and Viroid-like RNA Database is a compilation of 400 natural sequences from 36 different species (Lafontaine *et al.*, 1998). These sequences represent those published in journals and those available from GenBank and EMBL nucleotide sequence libraries. The four sections of the database are: (1) viroids, (2) satellite RNAs, (3) hepatitis delta virus RNAs, and (4) others. The curator of this database is Daniel Lafontaine from the Department of Biochemistry at the University of Sherbrooke in Quebec, Canada.

## 10.2.15   The urRNA Database

The urRNA Database is a compilation of sequences from small nuclear RNAs, e.g. U1–U8, U11, U12, U14, U48 and U49 (Zwieb, 1997). The curator of the urRNA Database is Christian Zwieb from the Department of Molecular Biology at the University of Texas Health Science Center in Tyler, TX, USA.

## 10.2.16   The RNA Modification Database

The RNA Modification Database lists 95 different examples of post-transcriptionally modified nucleosides from RNA (McCloskey & Crain, 1998). The information provided for each nucleoside includes type(s) of RNA in which are found: origin, phylogenetic distribution, chemical name and symbol, chemical structure and literature citations. The curators of the RNA Modification Database are Pamela Crain and James McCloskey from the Department of Medicinal Chemistry and Biochemistry at the University of Utah in Salt Lake City, UT, USA.

## 10.2.17   The Guide RNA Database

The Guide RNA Database lists 250 entries, sequences of small metabolically stable RNA molecules that participate in RNA-editing in kinetoplastid protozoa (Souza & Göringer, 1998). The curators of the Guide RNA Database are Augustine Souza and H. Ullrich Göringer from the Laboratory for Molecular Biology at the University of Munich in Martinsried, Germany.

## 10.2.18    A FastA-based compilation of higher plant mitochondrial tRNA genes

The compilation of higher plant mitochondrial tRNA genes contains 158 sequences derived from both literature data and analysis of the whole EMBL nucleotide sequence database library using the FASTA program (Sagliano *et al.*, 1998). The curator of this database is Luigi Ceci from the Department for Mitochondrial and Metabolic Energetics at Consiglio Nazionale delle Ricerche (CNR) in Bari, Italy.

## 10.2.19    tmRNA on-line resources

tmRNA (also known as 10Sa RNA) is so named for its tRNA-like and mRNA-like properties. Two resources currently exist to aid in the study of tmRNA-dependent protein tagging.

### 10.2.19.1    The tmRNA Website

The tmRNA Website includes tmRNA sequences from 37 species (Williams & Bartel, 1998). The curator of the tmRNA Website is Kelly Williams from the Whitehead Institute for Biomedical Research and Department of Biology at the Massachusetts Institute of Technology in Cambridge, MA, USA.

### 10.2.19.2    The tmRNA Database

Nineteen bacterial tmRNA sequences are listed in the tmRNA Database (tmRDB) (Zweib *et al.*, 1998). The curator of the tmRDB is Christian Zweib from the Department of Molecular Biology at the University of Texas Health Center in Tyler, TX, USA.

## 10.2.20    U-insertion/deletion Edited Sequence Database

The U-insertion/deletion Edited Sequence Database is a compilation of mito-chondrial genes and edited mRNAs from five kinetoplastid species. The database includes separate files with the DNA, mRNA (both edited and unedited), and predicted protein sequences (Simpson *et al.*, 1998). The database also contains genomic maps for each species and alignments of nuclear rRNA sequences for running phylogenetic programs. The curator of this database is

Larry Simpson from Howard Hughes Medical Institute at the University of California at Los Angeles, CA, USA.

## 10.2.21 The Database of Ribosomal Cross-links (DRC)

The Database of Ribosomal Cross-links (DRC) provides the collection of all published cross-linking data in the *E. coli* ribosome (Baranov *et al.*, 1998). The data include rRNA-rRNA crosslinks, rRNA-r-protein crosslinks, crosslinks between the r-proteins, between mRNA and the ribosome, tRNA and the ribosome, growing peptide and ribosome, factors and ribosome, and between ribosomal ligands. The curator of DRC is Pavel Baranov from the Max Planck Institute for Molecular Genetics in Berlin, Germany.

## 10.2.22 UTRdb: a specialized database of 5′ and 3′ untranslated regions of eukaryotic mRNAs

UTRdb, a specialized database of 5′ and 3′ untranslated sequences of eukaryotic mRNAs, is enriched with specialized information, such as sequence patterns implicated in functional roles for gene regulation and expression (Pesole *et al.*, 1998). The curator of UTRdb is Graziano Pesole at CNR in Bari, Italy.

## 10.3 TOOLS FOR ANALYSIS: RNA STRUCTURE AND PREDICTION

Examples of recent articles describing computational biology tools suitable for the analysis of RNA structure and prediction of RNA structure include:

- Structural analysis by energy dotplot of a large mRNA (Jacobson & Zuker, 1993)
- Automatic display of RNA secondary structures (Muller *et al.*, 1993)
- 'Well-determined' regions in RNA secondary structure prediction: analysis of small subunit ribosomal RNA (Zuker & Jacobson, 1995)
- Sequence walkers: a graphical method to display how binding proteins interact with DNA or RNA sequences (Schneider, 1997)
- XRNA 7.0, X Windows, RNA display and editing software (Weiser & Noller (*see Refs*); see also Gutell's (*See Refs*) RNA Secondary Structures WWW site)
- From sequences to shapes and back: a case study in RNA secondary structures (Schuster *et al.*, 1994)

- ESSA: an integrated and interactive computer tool for analysing RNA secondary structure (Chetouani *et al.*, 1997)
- RNA secondary structure as a reusable interface to biological information resources (Felciano *et al.*, 1997)
- Finding the most significant common sequence and structure motifs in a set of RNA sequences (Gorodkin *et al.*, 1997)
- Vienna RNA Package (public domain software) (Hofacker *et al.*, 1994)
- RAGA: RNA sequence alignment by genetic algorithm (Notredame *et al.*, 1997)
- Dynamic programming algorithm for the density of states of RNA secondary structures (Capal *et al.*, 1996)
- tRNAscan-SE (Lowe & Eddy, 1997)
- RNADraw (http://rnadraw.base8.se)
- MFOLD WWW server (http://www.ibc.wustl.edu/~zuker/rna/)
- Extinction coefficient calculation for DNA and RNA oligomers (http://paris.chem.yale.edu/extinct.html)
- TARD: Translational Active Regions Database (Novosibirsk) (http://benpc.bionet.nsc.ru/SRCG/Translation/index.html)
- Dynamic competition between alternative structures in viroid RNAs simulated by an RNA folding algorithm (Gultyaev *et al.*, 1998)

## 10.4  FUTURE DIRECTIONS

## 10.4.1  Catalogues

As the amount of network information related to RNA increases, the need for organization of this material has led to the construction of Web sites to provide centralized access to the information. Examples of on-line catalogues that provide extensive collections of software and include links to on-line resources related to RNA science include:

- RNA World (Sühnel, 1997)
- Image Library of Biological Macromolecules (Sühnel, 1996)
- INFOBIOGEN (e.g. Virgil: a database of rich links between GDB and GenBank; Achard & Barillot, 1998)
- The Nucleic Acid Database (Berman, *see Refs*)
- The Nucleic Acids Research database issue, published in January 1999, includes a report from Christian Burks tabulating and indexing on-line resources announced and updated in that journal (Roberts, 1998; Burks, 1999). The information is compiled into a list which includes links to both source Web sites and to on-line summaries describing the databases.

## 10.4.2 Genome projects

Examples of genome project Web sites that provide additional access to information related to RNA science include:

- TIGR: The Institute for Genome Research (http://www.tigr.org)
- KEGG: Kyoto Encyclopedia of Genes and Genomes (Kanehisa *see Refs*)
- The organelle genome database project (GOBASE) (Korab-Laskowska *et al.*, 1998)

## 10.4.3 The RNA Society

The RNA Society, which publishes the journal, *RNA*, was established in 1993 to facilitate sharing and dissemination of experimental results and emerging concepts in RNA research. The Society is multidisciplinary, representing the fields of biochemistry, biomedical sciences, chemistry, genetics, virology, molecular, evolutionary and structural biology, in relation to questions of the structure and function of RNA and of ribonucleoprotein complexes. The Society encompasses RNA research on the ribosome, the spliceosome, RNA viruses, and catalytic RNAs. The Society maintains two on-line resources: RNA Society Homepage (http://www.pitt.edu/~rna1/) and the *RNA* Homepage (http://www.cup.org/Journals/JNLSCAT/rRNA/rna.html).

The second and third annual meetings of the RNA Society included Bioinformatics workshops organized by Russ Altman. Altman provides on-line access to the proceedings of these workshops: RNA 97 Bioinformatics Workshop (Altman, 1997) and RNA 98 Informatics Workshop (Altman, 1998).

## 10.4.4 Educational resources

Four useful Web sites that provide a variety of RNA-based methods and tools include:

- Dartmouth Center for Biological & Biomedical Computing (Gross *see Refs*)
- NM (BIO502) RNA Structure and Prediction (Nelson & Istrail *see Refs*)
- Pedro's BioMolecular Research Tools (Coutinho *see Refs*)
- Cell and Molecular Biology on-line (http://www.cellbio.com)

Four recently published books describe and explain some of the computational biology tools found on the Internet:

- *Biocomputing: Informatics and Genome Projects* (Smith, 1994), which includes supercomputers, parallel processing, and genome projects (Smith et al., 1994)
- *Internet for the Molecular Biologist* (Swindell *et al.*, 1996), which includes sequence retrieval and analysis using electronic mail servers (Pillay, 1996)
- *Sequence Data Analysis Guidebook* (Swindell, 1997), which includes such topics as The Genetic Data Environment: a user modifiable and expandable multiple sequence analysis package (Eisen, 1997), The European Bioinformatics Institute: network service (Emmert *et al.*, 1994; Flores & Harper, 1997) and Primerselect: primer and probe design (Plasterer, 1997)
- *The Internet and the New Biology* (Peruski & Peruski, 1997)

## ACKNOWLEDGEMENTS

The chapter title originated from an RNA Society workshop organized by Russ Altman in 1997. The author is grateful to Franklin and Marshall College students who contributed to the review of RNA-related on-line resources: Robert M. Anthony, Sara E. Achenbach, Shannon R. Bowman, Kenneth D. Bromberg, Tiffany A. Freed, Ashley Grosvenor, Jyoti Gupta, Rashmi Khadilkar, Amer A. Khan, Julianna Kim, Melissa A. Kuehl, Mary L. Kupchella, Daniel J. Pedersen, Lee F. Richardson, Mark B. Schuster, Joshua P. Steinhaus, and Elizabeth A. Stewart. The Ribosomal RNA Mutation Database is supported by grants MCB 9315443 and MCB 9726951 from the National Science Foundation.

## REFERENCES

Achard F. & Barillot E. (1998) *Nucl. Acids Res.* **26**, 100–101. (http://www.infobiogen.fr/services/virgil/home.html)

Altman R.B. (1997) RNA 97 (http://www-smi.stanford.edu/people/altman/rna97.html)

Altman R.B. (1998) RNA 98 (http://www.wisc.edu/union/info/conf/rna/rna.html)

Baranov P.V., Sergiev P. V., Dontsova O.A., Bogdanov A. & Brimacombe R. (1998) *Nucl. Acids Res.* **26**, 187–189. (http://Ribosome.Genebee.MSU.SU/DRC)

Berman H. Nucleic Acid Database. (http://ndbserver.rutgers.edu)

Brown J.W. (1998) *Nucl. Acids Res.* **26**, 351–352. (http://jwbrown.mbio.ncsu.edu/RNAseP/home.html)

Burks C. (1999) *Nucl. Acids Res.* **27**, 1–9. (http://www.oup.co.uk/nar/Volume_27/issue_01/summary/gkc105_gm/.html)

Capal J., Hofacker I.L. & Stadler, P. F. (1996). In *Computer Science and Biology; Proceedings of the German Conference on Bioinformatics (GCB96)*, R. Hofestadt, T. Lengauer, M. Loffler & D. Schomburg (eds). University of Leipzig IMISE Report No. 1, pp. 184-186. Institute for Medical Informatics, Statistics and Epidemiology, Liebigstr. 27, 04103 Leipzig.

Chetouani F., Monestie P., Thebault P., Gaspin C. & Michot B. (1997) *Nucl. Acids Res.* **25**, 3514–3522.

Coutinho P.M. Pedro's BioMolecular Research Tools (http://www.public.iastate.edu/`pedro/research_tools.html)

Dalphin M.E., Brown C.M., Stockwell P.A. & Tate W.P. (1998) *Nucl. Acids Res.* **26**, 335–337. (http://biochem.otago.ac.nz:800/Transterm/homepage.html)

Damberger S.H. & Gutell R.R. (1994) *Nucl. Acids Res.* **22**, 3508–3510. (http://pundit.colorado.edu:8080/RNA/GRPI/introns.html)

de Rijk P., Caers A., van de Peer Y. & De Wachter R. (1998) *Nucl. Acids Res.* **26**, 183–186. (http://rrna.uia.ac.be/)

Eisen J.A. (1997). In *Sequence Data Analysis Guidebook*, S.R. Swindell (ed.) Humana Press Inc., Totowa, pp. 13–38.

Emmert D.B., Stoehr P.J., Stosser G. & Cameron, G.N. (1994) *Nucl. Acids Res.* **22**, 3445–3449.

Felciano R.M., Chen R.O. & Altman R.B. (1997) *Gene* **190**, GC59–GC70. (http://www-smi.stanford.edu/projects/helix/pubs/gene-combis-96/)

Flores T.P. & Harper R.A (1997). In *Sequence Data Analysis Guidebook*, S.R. Swindell (ed.) Humana Press Inc., Totowa, pp. 155–171.

Gorodkin J., Heyer L.J. & Stormo, G.D. (1997) *Nucl. Acids Res.* **25**, 3724–3732. (http://www.cbs.dtu.dk/gorodkin/appl/slogo.html)

Gross R.H. Center for Biological & Biomedical Computing. (http://www.dartmouth.edu/artsci/bio/cbbc/)

Gu J., Chen Y. & Reddy R. (1998) *Nucl. Acids Res.* **26**, 160–162. (http://mbcr.bcm.tmc.edu/smallRNA/smallrna.html)

Gultyaev A.P., van Batenburg F.M.D. & Pleij C.W.A. (1998) *J. Mol. Biol.* **276**, 43–56.

Gutell, R.R. RNA Secondary Structures WWW Site (http://pundit.colorado.EDU:8080/)

Gutell R.R. (1993) *Nucl. Acids Res.* **22**, 3051–3054. (http://pundit.colorado.edu:8080/RNA/16S/16s.html)

Gutell R.R., Gray M.W. & Schnare M.N. (1993) *Nucl. Acids Res.* **21**, 3055–3074. (http://pundit.colorado.edu:8080/RNA/23S/23s.html)

Hofacker I.L., Fontana W., Stadler P.F., Bonhoeffer L.S., Tacker M. & Schuster P. Vienna RNA Package (1994). (ftp://ftp.itc.univie.ac.at /pub/RNA/ViennaRNA-1.03, public domain software).

Jacobson A.B. & Zuker M. (1993) *J. Mol. Biol.* **233**, 261–269.

Kanehisa M. Kegg: Kyoto Encyclopedia of Genes and Genomes Release 4.0 (http://www.genome.ad.jp/kegg/)

Korab-Laskowska M., Rioux P., Brossard N., Littlejohn T.G., Gray M.W., Lang B.F. & Burger G. (1998) *Nucl. Acids Res.* **26**, 138–144.

Lafontaine D.A., Mercure S., Poisson V. & Perreault J.-P. (1998) *Nucl. Acids Res.* **26**, 190–191. (http://callisto.si.usherb.ca/~jpperra/indexZ.html)

Larsen N., Olsen G.J., Maidak B.L., McCaughey M.J., Overbeek R., Macke T.J., Marsh T.L. & Woese C.R. (1993) *Nucl. Acids Res.* 21, 3021–3023. (http://www. cme.msu.edu/RDP)

Larsen N., Samuelsson T. & Zwieb C. (1998) *Nucl. Acids Res.* 26, 177–178. (http://psyche.uthct.edu/dbs/SRPDB/SRPDB.html)

Lowe T. & Eddy S.R. (1997) *Nucl. Acids Res.* 25, 955–964. (http://genome.wustl.edu/ eddy/tRNAscan-SE/)

McCloskey J.A. & Crain P. (1998) *Nucl. Acids Res.* 26, 196–197. (http://www-medlib. med.utah/RNAmods/RNAmods.html)

Muller G., Gaspin C., Etienne A. & Westhof E. (1993) *Comput. Appl. Biosci.* 9, 551–561.

Nelson M. & Istrail S. RNA Structure and Prediction (BIO502) (http://algodones. unm.edu/~phraber/rna.html)

Notredame C., O'Brien E.A. & Higgins D.G. (1997) *Nucl. Acids Res.* 25, 4570–4580.

Park A.L. and Triman K.L. (1999) *Nucl. Acids Res.* 27, 3. (http://www.oup.co.uk/ nar/Volume_27/Issue_01/summary/nardb003.html)

Peruski L.F. & Peruski A.H. (1997). *The Internet and the New Biology.* American Society for Microbiology, Washington, DC.

Pesole G., Liuni S., Grillo G. & Saccone C. (1998) *Nucl. Acids Res.* 26, 192–195. (http://bio-www.ba.cnr.it:8000/srs5)

Pillay T.S. (1996) In *Internet for the Molecular Biologist*, Swindell S.R., Miller R.R. & Myers G.S.A. (eds). Horizon Scientific Press, Wymondham, pp. 59–72.

Plasterer T.N. (1997). In *Sequence Data Analysis Guidebook*, S.R. Swindell (ed.), Humana Press Inc., Totowa, pp. 291–302.

Roberts R.J. (1998) *Nucl. Acids Res.* 26, i (Editorial).

Sagliano A., Volpicella M., Gallerani R. & Ceci L.R. (1998) *Nucl. Acids Res.* 26, 154–155. (http://www.ebi.ac.uk/service)

Schneider T.D. (1997) *Nucl. Acids Res.* 25, 4408–4415. (http://www-lmmb.ncifcrf. gov/~toms/)

Schuster P., Fontana W., Stadler P.F. & Hofacker I.L. (1994) *Proc. Roy. Soc. Lond. B.* 255, 279–284.

Simpson L., Wang S.H., Thiemann O.H., Alfonzo J.D., Maslov D.A. & Avila H.A. (1998) *Nucl. Acids Res.* 26, 170–176. (http://www.lifesci.ucla.edu/RNA/trypanosome/ database.html)

Smith D.W. (ed.) (1994) *Biocomputing: Informatics and Genome Projects.* Academic Press, Inc., San Diego.

Smith D.W., Jorgensen J., Greenberg J.P., Keller J., Rogers J., Gerner H.R. & Eyck L.T. (1994). In *Biocomputing: Informatics and Genome Projects*, D.W. Smith (ed.) Academic Press, Inc., San Diego, pp. 51–86.

Souza A.E. & Göringer, H.U. (1998) *Nucl. Acids Res.* 26, 168–169. (http://www. biochem.mpg.de/~goeringe/)

Specht T., Wolters J. & Erdmann V.A. (1991) *Nucl. Acids Res.* 19, 2189–2191.

Sprinzl M., Horn C., Brown M., Ioudovitch A. & Steinberg S. (1998) *Nucl. Acids Res.* 26, 148–153. (http://www.uni-bayreuth.de/departments/biochemie/trna/)

Sühnel J. RNA World (http://www.imb-jena.de/RNA.html)

Sühnel J. (1996) *Computer Applications in the Biosciences* 12, 227–229. (http://www. imb-jena.de/IMAGE.html)

Sühnel J. (1997) *Trends in Genetics* 12, 206–207.

Swindell S.R. (ed.) (1997). *Sequence Data Analysis Guidebook*. Humana Press Inc., Totowa.

Swindell S.R., Miller R.R. & Myers G.S.A. (eds.) (1996). *Internet for the Molecular Biologist*. Horizon Scientific Press, Wymondham.

Szymanski M., Specht T., Barciszewska M.Z., Barciszewski J. & Erdmann V.A. (1998) *Nucl. Acids Res.* 26, 156–159. (http://www.chemie.fu-berlin.de/fb_chemie/agerdmann/5S_rRNA.html or http://rose.man.poznan.pl/5SData/5SRNA.html)

Triman K., Peister A. & Goel R.A. (1998) *Nucl. Acids Res.* 26, 280–284. (http://ribosome.fandm.edu or http://www.fandm.edu/Departments/Biology/Databases/RNA.html)

van de Peer Y., Caers A., De Rijk P. & De Wachter R. (1998) *Nucl. Acids Res.* 26, 179–182. (http://rrna.uia.ac.be/ssu)

Weiser B. & Noller, H.F. XRNA 7.0, X Windows, RNA display and editing software. (ftp://fangio.ucsc.edu/pub/XRNA/)

Williams K.P. & Bartel D.P. (1998) *Nucl. Acids Res.* 26, 163–165. (http://www.wi.mit.edu/bartel/tmRNA/home)

Wu C. & Shivakumar S. (1994) *Nucl. Acids Res.* 22, 4291–4299. (http://diana.uthct.edu/~nih/cans/gencans_rdp.html)

Zuker M. & Jacobson A.B. (1995) *Nucl. Acids Res.* 23, 2792–2798.

Zwieb C. (1997) *Nucl. Acids Res.* 25, 102–103. (http://psyche.uthct.edu/dbs/uRNADB/uRNADB.html)

Zwieb C., Larsen N. & Wower J. (1998) *Nucl. Acids Res.* 26, 166–167. (http://www.uthct.edu/tmRDB/tmRDB.html)

# 11 | Predicting the Evolution, Structure and Function of Proteins from Sequence Information

*Chris P. Ponting[1] & D.J. Blake[2]*

[1]Oxford Centre for Molecular Sciences, University of Oxford, Oxford, UK
[2]University of Oxford Genetics Unit, Department of Biochemistry, Oxford, UK

## 11.1  INTRODUCTION

The predictive potential of protein sequence alignment in biology is unrivalled. Annotations of gene sequences deposited daily in databases testify to the considerable faith biologists place in the prediction of evolution, structure and function from sequence alone. The majority of novel proteins are annotated as possessing particular functions without recourse to experiment, simply due to significant (and sometimes insignificant) sequence similarities with homologues of experimentally derived functions. However, it is not apparent what degree of confidence should be placed in these predictions. For example, does knowledge of the function of a nematode worm protein shed light on the function of a human homologue, given that the anatomies of the organisms differ so greatly. Addressing this question requires consideration of what is meant by 'function'. Should this term be applied differently to structures with contrasting dimensions, such as amino acids or proteins, or cells or organisms? In this chapter, we discuss descriptions of protein function that are commonly used by sequence annotators and view, from an evolutionary perspective, the consequences of past genetic modifications that have resulted in the persistence of ancient genes up to the present day.

## 11.2  PROTEIN EVOLUTION AND FUNCTION

Novel molecular and cellular functions are acquired by organisms owing to changes in their genome sequences and may be positively selected. Conversely,

*Genetics Databases*
ISBN 0-12-101625-0

genome mutation can cause molecular and cellular dysfunction that results in disease and negative selection.[1] Consequently, it is important to relate the prediction of function to the consequences of evolution. Central to this is the concept of homology. Homologues are genes whose sequences often (but not always) share significant similarities and that represent modern versions of an ancient gene that at some moment in its past underwent gene duplication. Assessing the significance of sequence similarities is discussed elsewhere in this volume – additional reviews include Bork & Gibson (1996) and Ponting & Birney (1998). Since the duplication event each copy is considered to have mutated independently such that levels of pairwise sequence identity have eroded from the initial 100% value. It should be noted that this definition of homology renders meaningless the phrase 'percentage homology'; in its place, the term 'pairwise percentage sequence identity' should be used. Importantly, since homology is defined simply in terms of evolutionary relationships, it is not meant that homologues always possess identical functions (cf. Murzin, 1993).

Gene homologues may arise from speciation or from gene duplication within a genome (Ohno, 1970). Single homologues from different organisms that are most similar in sequence and therefore form a distinct branch of a phylogenetic tree are usually thought to have arisen from speciation events, and are termed 'orthologues' (Fitch, 1970). Orthologues have been defined as possessing identical functions (e.g. Doolittle, 1998; Bork & Koonin, 1998). However, it may be prudent to define orthology as an evolutionary relationship since it is not apparent how to quantify functionality in different organisms (Miklos & Rubin, 1996). It is unclear whether the presence of orthologues in a variety of diverse organisms always reflects an equivalence of their functions. However the identification of only 34 obvious orthologues common to the first nine known bacterial genomes (Huynen & Bork, 1998) suggests otherwise, since these cellular organisms are expected to possess many common housekeeper functions. Nevertheless, considerable (> 50%) pairwise sequence identity for a pair of orthologues from two closely related species is usually an excellent predictor of common functionality.

Homologues present in the same organism are thought to have arisen from past intra-genome gene duplication and are termed 'paralogues' (Fitch, 1970). This is of particular relevance to the comparison of vertebrate with inverte-

---

[1] However, mutations do persist within a population when they provide a selective advantage. Perhaps the most well known example of this is the incidence of haemoglobin S in populations of sub-Saharan Africa, the Middle East, India and parts of the Mediterranean (reviewed by Clegg, 1997). The mutation of Gln6→Val in β-globin results in sickle cell anaemia in the homozygous state. Heterozygotes for the mutation are clinically benign but are resistant to infection by the malaria parasite *Plasmodium falciparum*. Thus, the glutamate to valine substitution imparts a selective advantage to heterozygote carriers and is therefore maintained within the population even though the disease affects homozygotes.

brate sequences since two separate genome duplication events are thought to have occurred late in the history of vertebrates (Gehring & Hiromi, 1986; Holland and Garcia-Fernöndez, 1996). This results in vertebrate genomes containing a maximum of four non-identical versions of a gene (tetralogues; Spring, 1997) that are equally similar to a single homologous version contained in invertebrate genomes.

Does this genetic redundancy reflect a redundancy in protein function? If so, it might be expected that the majority of gene duplications might have been lost soon after their creation owing to the acquisition of null mutations. The half-life for dispensable vertebrate genes has been estimated as only 50 million years (Kimura, 1983) which is the approximate time since divergence of rat and mouse or cat and dog lineages (Kumar & Hedges, 1998). On one hand, positional cloning of genes mutated in disease that possess paralogues (Mushegian *et al.*, 1997) indicates that some paralogues are functionally non-redundant. Conversely, knockout of single paralogues can result in the lack of an observable phenotype (Manley & Capecchi, 1997; Teglund *et al.*, 1998) that is likely to be due to functional redundancy. In this regard it should be noted that redundancy of molecular function is not always equivalent to redundancy of cellular function. For example, the cellular roles of two sequence-identical genes, whose expression patterns are non-overlapping, would be nonredundant since one paralogue would be unable to compensate for another if either gene is mutated.

Preservation of multiple paralogues in vertebrate genomes appears to have occurred because of comparable rates of gene loss and of functional divergence following genome duplication (Nadeau & Sankoff, 1997). Evolutionary models that only consider persistence of paralogues due to the acquisition of novel and eventually essential functions (Ohno, 1970) cannot account for these similar rates. Nadeau & Sankoff (1997) account for persistence of paralogues by proposing a high rate of functional divergence for one of the duplicated genes via the accumulation of greater numbers of neutral over disadvantageous mutations. Gibson & Spring (1998) have advanced an hypothesis that paralogues containing multiple domains (see Section 11.4) diverge and are eliminated only slowly since these are less tolerant of deleterious point mutations than are single domain proteins.

Cellular life has evolved intricate systems of functionally synergistic molecules that act within pathways, networks and cycles to promote the survival and duplication of genomic DNA. Sequencing of eubacterial and archaeal genomes has demonstrated that such systems have persisted since the earliest forms of cellular life. However, in several cases, molecules that are not orthologues or even homologues appear to perform similar roles as key participants in a pathway, network or cycle: this has been termed 'non-orthologous gene displacement' (Koonin *et al.*, 1996). This suggests that multiple invention of function is relatively commonplace. Another way for an organism to acquire molecular function is for it to steal DNA from another source. Evidence for such 'lateral' or 'horizontal' transfer of genes between prokaryotes is compelling (reviewed in Doolittle, 1998; Doolittle & Logsdon, 1998).

## 11.3  PROTEIN STRUCTURE AND FUNCTION

The unit of protein structure that has been most conserved throughout evolution is the domain. The domain is a compact arrangement of secondary structures ($\alpha$-helices and/or $\beta$-strands) connected by 'linker' polypeptides, that usually folds independently and possesses a relatively hydrophobic core (Janin & Chothia, 1985). Domains may not be subdivided into other domains. Domains are distinct from motifs that are usually defined as small, highly sequence-similar regions of domains (Bork & Koonin, 1996) or else nonglobular structures such as transmembrane helices or coiled coils or signal peptides. Experience gained from protein structure determination throughout recent decades demonstrates that domains possessing similar sequences also possess similar folds, leading to the inference that such domains are homologues (Doolittle, 1995). Some homologous domain sequences have diverged substantially beyond the level at which homology can be predicted reliably. However, from the tertiary structures of these domains it is often seen that their folds are conserved even when their sequences are not. Thus, predicted homology from sequence similarity is an excellent predictor of structure similarity.

In a small number of instances homologues have been demonstrated to possess dissimilarities in structure (Russell & Ponting, 1998). Domain homologues are known to differ owing to insertions (Russell, 1994), circular permutations (Lindqvist & Schneider, 1997), disulphide swapping (Benitez *et al.*, 1997) and secondary structure exchange (Fletterick & Bajan, 1995). The most common of these is insertion. For example, the thioredoxin fold is adopted by a variety of enzymes that interact with cysteine-containing substrates (Martin, 1995). This fold is prone to insertions of secondary structures or even entire domains, either preceding, intervening or succeeding the thioredoxin-like scaffold.

In the absence of domain insertion, multidomain proteins consist of tandem arrays of homologous domains (tandem repeats) and/or nonhomologous domains (Heringa & Taylor, 1997). Domains can be loosely associated like 'beads-on-a-string' or intimately associated, with domain–domain interactions (see Section 11.5.2).

The concept of homology should be applied principally to domains rather than multidomain proteins. At the level of domains, homology can be inferred by induction if domain *A* is a homologue of domain *B*, and domain *B* is a homologue of domain *C*, then domains *A* and *C* are also homologues. However, at the level of multidomain proteins, homology cannot be inferred by induction if protein $X$ is a fusion of two nonhomologous domains, each of which is a homologue of single domain proteins $Y$ and $Z$, then $Y$ and $Z$ are not homologues. This often results in errors in gene annotations in databases (Galperin & Koonin, 1998). For example, significant similarity of a sequence with the Src protein kinase sequence might lead to its annotation as a kinase, even though the similarity is apparent only within the regulatory Src homology 2 (SH2) domain.

Similarities in domain function are usually apparent among the majority of domain homologues. For example, SH2 domains mediate protein–protein interactions via their interaction with phosphotyrosine-containing polypeptides (Kuriyan & Cowburn, 1997). Such interactions may be intramolecular, as demonstrated in the crystal structures of Src paralogues (Xu *et al.*, 1997; Sicheri *et al.*, 1997, see Section 11.5.2), and/or intermolecular as in adapter molecules. This does not preclude further interactions with other substrates that do not involve the phosphotyrosine-binding site (Park *et al.*, 1995). Consequently, the prediction of function of a multidomain protein is a three-step procedure: (1) prediction of the functions of individual domains, (2) prediction of the protein function from its modular architecture, and (3) prediction of the involvement of the protein in particular pathways or networks. Each of these steps requires analogies to be drawn to experimentally derived functions of: (1) domain homologues, (2) protein homologues (i.e. proteins with identical or near-identical arrangements of domains), and (3) analogous pathways or networks.

## 11.4 FUNCTIONS OF ENZYMATIC AND REGULATORY DOMAINS

There are six main classes of enzymes: oxidoreductases, transferases, hydrolases, lyases, isomerases, and ligases (Dixon & Webb, 1979, see http://www.expasy.ch/sprot/enzyme.html); each class contains considerable numbers of nonhomologous enzyme families. Families usually contain representatives that perform diverse molecular and cellular functions. For example, the AAA family of ATPases perform cellular functions ranging from cell cycle regulation to transcriptional activation to endocytosis (Confalonieri & Duguet, 1995). In this, and the majority of cases, enzymes with similar cellular functions form a distinct branch of the family's phylogenetic tree. However, in a minority of cases that result from unusual selective pressures on organisms, function does not partition with sequence similarity, as with the Crotalinae snake venom gland serine proteases (Deshimaru *et al.*, 1996).

Enzyme structures may contain one or more domains contributed by monomers or multimers. Amino acids that participate in catalysis are usually partially buried in a cleft or depression, and, in general, are the most conserved among families of enzyme homologues. Domains that do not possess enzymatic activity are usually less conserved and are smaller than enzymes with similar ancestry. These regulatory domains often act by binding cofactors that serve to localize the protein to a cellular compartment or to initiate conformational change. Such domains may bind one or more of nucleic acids, lipids, proteins, prosthetic groups and ions, rendering their regulation of function exquisitely intricate. Consequently, prediction of catalysis or binding function requires conservation of each of the amino acid determinants known to be required for that function. For example, the presence or absence of three basic residues differentiates those among a family of Ras-binding domains (Ponting & Benjamin, 1996) that bind Ras$^{GTP}$ (Vavvas et al., 1998; Shibatohge et al., 1998; Huang et al., 1998) and those that do not (Kalhammer et al., 1997).

The distinction between enzymatic domains and regulatory domains is partly blurred by recent identifications of enzyme homologues that possess substitutions of catalytic residues. These include apparently inactive homologues of protein kinases (Wilks, 1989), phosphatases (Wishart et al., 1995; Haynie & Ponting, 1996), serine proteases (Kurosky et al., 1980; Nakamura et al., 1989), nucleases (Ponting, 1997), ubiquitin-conjugating enzymes (Koonin & Abagyan, 1997; Ponting et al., 1997), ureases (Holm & Sander, 1997), caspases (Han et al., 1997) and guanylate kinases (Kistner et al., 1995). This phenomenon is not exclusive to enzymes, as demonstrated by ovalbumin which, unlike its serpin homologues, is inactive as an inhibitor of serine proteases (Wright, 1984).

## 11.5   CLASSICAL GENETICS AND PROTEIN FUNCTION

Classical genetics has concerned itself with the analysis of protein function through mutation. Mutations generated within a gene of interest can be seen to abolish or perturb the function of a protein. Perhaps the simplest genetic model for studying protein function is gene deletion. Construction of a null allele in the genome of an organism can be achieved using a variety of direct and indirect manipulations. These mutations, when in their homozygous state, provide us with an organism that lacks the protein of interest. Intuitively, the analysis of the mutant phenotype should ultimately lead to a description of the organismal function of a protein. For example, mutations in the *tinman* gene of *Drosophila* result in the absence of the heart (Azpiazu & Frasch, 1993). The *tinman* gene encodes a homeodomain-containing transcription factor that is required for the formation of the visceral mesoderm, heart progenitors and somatic muscle precursors. *Tinman* binds to the promoters of certain muscle-expressed transcription factors and protein coding genes whose products are required for muscle formation (Gajewski *et al.*, 1997). Thus, null mutations in the *tinman* gene ultimately prevent the temporal transcription of several proteins, leading to the absence of the heart and other mesodermal structures.

Other types of mutation can be used to explore protein function. For example dominant negative mutants when over-expressed can interfere with protein function. These mutants may be used to perturb the assembly of oligomeric, often transcriptional or membrane-associated, protein complexes. For instance, a substitution of Arg271→Trp in one allele of the Pit-1 transcription factor allows this mutant protein to bind normally to DNA but blocks transcription (Radovick *et al.*, 1992). This dominant-inhibition of transcription results in a combined pituitary hormone deficiency.

Gain-of-function mutations also provide information about protein function. Perhaps one of the best examples of a gain-of-function mutation is in the $\alpha_1$-antitrypsin gene. Mutations in this gene lead to hereditary emphysema and liver disease. However, an amino acid substitution, Met358→Arg, alters the specificity of this serine protease inhibitor from its physiological substrate, neutrophil elastase, to thrombin (Owen *et al.*, 1983). This mutation causes a bleeding disorder that is quite distinct from the lung disease typical of $\alpha_1$-antitrypsin deficiency.

While genetics can be used as a powerful tool to examine protein function, genetic analysis of some proteins often is confusing. This is of particular relevance to vertebrate proteins since, typically, these are multidomain in character and are members of large families of paralogues and/or domain homologues. Here we illustrate our previous discussion by reviewing the relationship between prediction of domain architecture from sequence and prediction of function for two molecules, dystrophin and Src, that are examples of multidomain proteins that possess paralogues. Each of these is associated with a well-studied disease state and each contains domains whose folds are known from structure determination. In contrast to the wealth of knowledge on Src and dystrophin sequences and structures, little is known about their cellular roles. Moreover, regulation of Src function was not fully apparent until determination of its tertiary structure showed substantial co-operativity of its constituent domains' functions.

## 11.5.1 Dystrophin

Dystrophin is the product of the Duchenne muscular dystrophy (DMD) gene and was one of the first human genetic disease genes to be identified by positional cloning (reviewed in Blake & Davies, 1997). In its most severe form, DMD is a lethal condition. Boys suffering from the disease die in their early twenties from pulmonary or cardiac insufficiency. It is important to note that although models of DMD exist in other mammalian species, such as mice, these incompletely resemble the disease in humans. For example, the *mdx* mouse has a point mutation in the dystrophin gene that effectively abolishes production of any functional dystrophin yet still lives a normal life span. Thus, the organismal functions of human and mouse dystrophin orthologues are not always equivalent. This apparent inequality may merely reflect differences between species, such as the difference in life span between humans and mice or may result from an alteration in the protein function gained or lost through evolution.

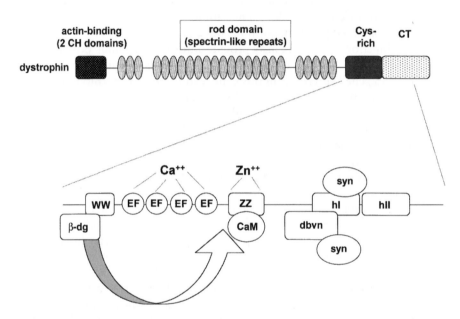

**Figure 11.1**    Schematic representation of the domains in dystrophin. The *N*-terminus of dystrophin binds to F-actin through two CH domains. The majority of the dystrophin molecule is the so-called rod domain that is composed of 24 spectrin-like repeats (ovals) interrupted by four proline-rich hinge regions. The C-terminus of dystrophin can be divided into cysteine-rich and C-terminal (CT) regions. The Cys-rich region is preceded by the WW domain, that is in part, required to bind to β-dystroglycan. The Cys-rich region is composed of four EF-hand motifs that are not known to bind $Ca^{2+}$, and a ZZ domain that is required for calmodulin (CaM) binding. The C-terminus of dystrophin contains two regions of predicted coiled-coil, helix I (hI) and helix II (hII). This region mediates the interaction between dystrophin and the cytoplasmic components of the dystrophin-associated protein complex. Both dystrobrevin (dbvn) and syntrophin (syn) bind to similar regions in helix I. The interaction between dystrophin and the dystrophin-related protein dystrobrevin involves the coiled-coil domains of each protein. In addition, syntrophin can also bind to dystrobrevin, thereby forming a quaternary complex at the C-terminus of dystrophin.

Dystrophin is encoded by 79 exons that span almost 2.5 Mb of the human X-chromosome. It is a large modular protein (Fig. 11.1) containing an *N*-terminal pair of calponin homology (CH) domains (Castresana & Saraste, 1995), 24 spectrin-homologous coiled-coil repeats (Winder *et al.*, 1995), and a *C*-terminal region with WW (Bork & Sudol, 1994) and ZZ (Ponting *et al.*, 1996) domains and two coiled-coil motifs (Blake *et al.*, 1995). Utrophin is a paralogue of dystrophin with which it shares an almost identical domain architecture (Tinsley *et al.*, 1992); dystrophin and utrophin appear to have arisen from gene duplication during vertebrate evolution (Roberts & Bobrow, 1998). Utrophin-deficient mice have mild neuromuscular junction abnormalities but normal muscle function (Deconinck *et al.*, 1997a; Grady *et al.*, 1997a). Mice with mutations in both dystrophin and utrophin genes are severely affected and die from respiratory failure within 10 weeks of birth (Deconinck *et al.*, 1997b; Grady *et al.*, 1997b). The muscle disease in these double mutant mice is very similar to that observed in DMD patients and suggests that, in mice, utrophin can partially compensate for the loss of dystrophin. Thus, these paralogues have non-identical yet overlapping cellular and organismal functions that could appear to be genetically redundant.

Despite extensive investigation, no definitive molecular function for dystrophin or utrophin has yet been described. One hypothesis is that dystrophin plays a role in maintaining the integrity of the muscle membrane during the stresses of contraction and relaxation. Indeed, dystrophin is well qualified to perform such a structural role as it contains 24 triple-helical spectrin-like repeats and, like spectrin, is a member of a membrane-associated complex. Mutations in the α- and β-spectrin genes cause hereditary elliptocytosis and sphereocytosis, diseases characterized by abnormally shaped erythrocytes that are mechanically fragile. Spectrin-deficient erythrocytes are weaker than normal erythrocytes and are prone to haemolysis, leading to an anaemic condition. Thus, obvious parallels exist between spectrin in erythrocytes and dystrophin in muscle: both proteins are integral components of a membrane-associated complex and loss of this complex results in increase in membrane fragility.

The original report of the dystrophin sequence (Koenig *et al.*, 1988) predicted, on the basis of α-actinin homology, actin-binding, homodimerization and $Ca^{2+}$-binding functions for its CH, spectrin and EF-hand domains. Only one of these functions, that of CH domain-mediated actin-binding, has been demonstrated (reviewed in Winder *et al.*, 1995). Moreover, the subsequent prediction of WW and ZZ domains, homologous to polyproline-binding signalling (Bork & Sudol, 1994) and transcriptional adapter/coactivator proteins (Ponting *et al.*, 1996) in the *C*-termini of dystrophin and utrophin did not lead directly to insight into their functions. The complete WW domain of dystrophin is not absolutely required for binding of the proline-rich region of β-dystroglycan (Jung *et al.*, 1995) and a calmodulin-binding function of the ZZ domain (Anderson *et al.*, 1996) was not anticipated. ZZ domains represent one family of putative zinc fingers. These are often predicted from sequence to be DNA-binding motifs. However, many zinc finger-containing proteins are not nuclear and bind to proteins and not DNA (Mackay & Crossley, 1998). Clearly, the ZZ domain is important for the cellular function of dystrophin since transgenic mice that over-express a mutant dystrophin protein lacking the ZZ domain have muscle disease and fail to localize the dystrophin–protein complex to the muscle membrane (Rafael *et al.*, 1996). However, identification of the domain failed to allow prediction of its function.

In contrast, a sequence-based prediction that the C-terminal coiled-coil regions of dystrophin and the dystrophin-related protein β-dystrobrevin interact (Blake *et al.*, 1995) has been corroborated experimentally (Sadoulet-Puccio *et al.*, 1997). Syntrophin also binds directly to dystrobrevin via its coiled-coil region forming a quaternary complex (Peters *et al.*, 1997; Fig. 11.1). The physiological role of these interactions is unknown. Transgenic mice that lack the syntrophin/dystrobrevin-binding site and also the second coiled-coil helix are phenotypically normal and have a normal location of the DPC including syntrophin and dystrobrevin (Rafael *et al.*, 1996; Peters *et al.*, 1997). One possibility is that in the absence of the dystrophin's coiled-coil region syntrophin and dystrobrevin bind to its spectrin-like coiled-coils.

In summary, although a great deal is known about the pathology of DMD and the interactions of dystrophin with the muscle membrane, dystrophin's cellular role remains elusive. Although dystrophin is thought to play a structural role in contracting muscle, dystrophin-deficiency also affects cognitive processes in the brain (Lidov, 1996). Thus, dystrophin may have a multitude of functions, some of which may be mechanical and others that may relate to anchoring signalling molecules or membrane-associated complexes.

## 11.5.2   The Src tyrosine kinase

Viral Src was initially identified as the oncogenic component of the Rous avian sarcoma retrovirus. The cellular counterpart of this viral oncogene, the proto-oncogene, like the viral protein, has the ability to transform cells and form tumours when mutated. The Src family of nonreceptor tyrosine kinases consists of nine paralogues that share a common domain architecture, containing Src homology 2 and 3 (SH2 and SH3) domains and a C-terminal tyrosine kinase domain (Fig. 11.2). Although Src has been implicated in a wide variety of signal transduction pathways, targeted disruption of the Src gene results in a surprisingly mild phenotype. Src-deficient mice have mild osteopetrosis that leads to abnormal dentition (Soriano *et al.*, 1991).

Targeted disruption of other members of the Src family of kinases often results in mild, non-overlapping phenotypes. For instance, Fyn-deficient mice have abnormalities in long-term potentiation, the molecular mechanism

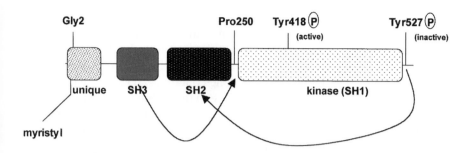

**Figure 11.2** Schematic representation of the Src tyrosine kinase. Src contains four domains separated by peptide linkers. The *N*-terminal domain contains a myristylation site. In its inactive conformation the Src SH3 domain intramolecularly binds Pro250 whereas the SH2 domain binds the C-terminal tail via phosphorylated Tyr527. These interactions together with associations between the SH2-kinase linker and the *N*-terminal lobe of the kinase domain result in Src's inactive conformation. It is thought that, upon activation, the C-terminal tail becomes dephosphorylated severing its interaction with the SH2 domain and that this exposes the kinase catalytic site, resulting in autophosphorylation of Tyr418. Dissociation of intramolecular interactions allows Src to interact with its various substrates.

thought to underpin memory (Grant *et al.*, 1992) whereas Yes-deficient mice lack any observable phenotype (Stein *et al.*, 1994). More severe phenotypes are revealed when double mutants are created. Mice with deficiencies of both Src and Fyn, and Src and Yes die perinatally, suggesting that the Src family members partly compensate for each other (Stein *et al.*, 1994).

How do these mutant phenotypes relate to the proteins' cellular and molecular functions? It is important to note that the conclusions drawn from the analysis of targeted mice may be unreliable. As previously mentioned, several Src paralogues appear to compensate for each other. It could therefore be concluded that these have overlapping substrates and/or participate in the same signalling cascades. However, in contrast to the mild phenotype observed in mice with gene deletions, missense mutations of these kinases can cause cellular transformation ultimately leading to tumour formation. These mutations alter the regulation, specificity or association of each kinase and therefore differ from the null mutants that effectively produce no protein.

Recently solved structures of the inactive states of Src and other Src-family kinases have revealed the intricate associations between the different domains of Src and explain how this kinase is regulated (Xu *et al.*, 1997; Sicheri *et al.*, 1997; Williams *et al.*, 1998). Src contains three well-characterized domains, the SH2 domain, the SH3 domain and the kinase domain (Fig. 11.2). Membrane localization of Src is required for Src-mediated signal transduction and is achieved by myristoylation at Gly2. Other Src family kinases are also post-translationally modified at their N-termini by palmitoylation. Src kinase activity is regulated by two intramolecular interactions: that between the SH2 domain and a C-terminal phosphorylated tyrosine (Tyr527 in chicken Src) and that between the SH3 domain and a proline-containing SH2-kinase linker peptide. Consequently, the SH2 and SH3 domains loosely associate with the kinase domain, locking it into a relatively inactive conformation. Src becomes activated when Tyr527 is dephosphorylated, or when a protein binds to the SH2 or SH3 domain, or due to interaction with an allosteric activator such as cdc2/cyclin B. Activated Src adopts an 'open' conformation allowing the SH2 and SH3 domains to interact with other proteins. Although the 'closed' conformation of inactive Src is favoured, these interactions can occur since the open and closed conformations are in dynamic equilibrium.

The intramolecular interactions that maintain Src in its inactive state explain how several mutations can cause constitutive activation of the kinase. For example, mutating Tyr527→Phe, or critical residues in the SH2 domain, result in kinase activation and cellular transformation (Hunter, 1987; Xu et al., 1997; Sicheri et al., 1997; and references therein). It is clear from Src structures that these mutations effectively abolish the association between the SH2 domain and the protein's C-terminal tail, thus facilitating interaction of SH2 and SH3 domains with other proteins. It is not yet fully apparent how adoption of the open conformation leads to an active kinase function. Thus, although multiple binding functions of Src domains are essential requirements for the physiological regulation of Src's functions, as demonstrated by knockout studies, the organismal function of Src is not.

## 11.6  SUMMARY

Protein function annotation by homology is currently of poor quality. Generally, little attention is paid to the multidomain architectures of proteins. More importantly, the essential distinctions between descriptions of function at the amino acid, domain, multiprotein system, cell and organism levels often are not understood. In a few years a wealth of information relating to genomes, proteomes, protein structures and gene knockouts should be available. It will be essential that predictions of function make optimal use of these data.

## REFERENCES

Anderson J.T., Rogers R.P. & Jarrett H.W. (1996) J. Biol. Chem. 271, 6605–6610.
Azpiazu N. & Frasch M. (1993) Genes Dev. 7, 1325–1340.

Benitez B.A.S., Hunter M.J., Meininger D.P. & Komives E.A. (1997) *J. Mol. Biol.* 273, 913–926.

Blake D.J., Tinsley J. M., Davies K. E., Knight A.E., Winder S. J. & Kendrick-Jones J. (1995) *Trends Biochem. Sci.* 20, 133–135.

Blake D.J. & Davies K.E. (1997) In *Protein Dysfunction and Human Genetic Disease*, Swallow D.M. & Edwards Y.H. (eds). BIOS Scientific Publishers, Oxford, pp. 219–241.

Bork P. & Gibson T.J. (1996) *Meth. Enzymol.* 266, 162–184.

Bork P. & Koonin E.V. (1996) *Curr. Opin. Struct. Biol.* 6, 366–376.

Bork P. & Koonin E.V. (1998) *Nature Genet.* 18, 313–318.

Bork P. & Sudol M. (1994) *Trends Biochem. Sci.* 19, 531–533.

Castresana J. & Saraste M. (1995) *FEBS Lett.* 374, 149–151.

Clegg J.B. (1997) In *Haemoglobin. Protein Dysfunction and Human Genetic Disease*, Swallow D.M. & Edwards Y.H. (eds). BIOS Scientific Publishers, Oxford, pp. 15–31.

Confalonieri F. & Duguet M. (1995) *BioEssays* 17, 639–650.

Deconinck A.E., Potter A.C., Tinsley J.M., Wood S.J., Vater R., Young C., Metzinger L., Vincent A., Slater C.R. & Davies K.E. (1997a) *J. Cell Biol.* 136, 883–894.

Deconinck A.E., Rafael J.A., Skinner J.A., Brown S.C., Potter A.C., Metzinger L., Watt D.J., Dickson J.G., Tinsley J.M. & Davies K.E. (1997b) *Cell* 90, 717–727.

Deshimaru M., Ogawa T., Nakashima K., Nobuhisa I., Chijiwa T., Shimohigashi Y., Fukumaki Y., Niwa M., Yamashina I., Hattori S. & Ohno M. (1996) *FEBS Lett.* 397, 83–88.

Dixon M. & Webb E.C. (1979) *Enzymes*. Longman Group, London.

Doolittle R.F. (1995) *Annu. Rev. Biochem.* 64, 287–314.

Doolittle R.F. (1998) *Nature* 392, 339–342.

Doolittle W.F. & Logsdon J.M. Jr. (1998) *Curr. Biol.* 8, R209–R211.

Fitch W.M. (1970) *Syst. Zool.* 19, 99–113.

Fletterick R.J. & Bazan J.F. (1995) *Nature Struct. Biol.* 2, 721–723.

Gajewski K., Kim Y., Lee Y. M., Olson E. N. & Schulz R. A. (1997) *EMBO J.* 16, 512–515.

Galperin M.Y. & Koonin E.V. (1998) *In Silico Biol.* 1, 0007. (http://www.bioinfo.de/isb/1998/01/0007/)

Gehring W.J. & Hiromi Y. (1986) *Annu. Rev. Genet.* 20, 147–173.

Gibson T.J. & Spring J. (1998) *Trends Genet.* 14, 46–49.

Grady R.M., Merlie J.P. & Sanes J.R. (1997a) *J. Cell Biol.* 136, 871–882.

Grady R. M., Teng H., Nichol M., Cunningham J. C., Wilkinson R. S. & Sanes J.R. (1997b) *Cell* 90, 729–738.

Grant S.G., O'Dell T.J., Karl K.A., Stein P.L., Soriano P. & Kandel E.R. (1992) *Science* 258, 1903–1910.

Han D.K.M., Chaudhary P.M., Wright M.E., Friedman C., Trask B.J., Riedel R.T., Baskin D.G., Schwartz S.M. & Hood S. (1997) *Proc. Natl. Acad. Sci. USA* 94, 11333–11338.

Haynie D.T. & Ponting C.P. (1996) *Prot. Sci.* 5, 2643–2646.

Heringa J. & Taylor W.R. (1997) *Curr. Opin. Struct. Biol.* 7, 416–421.

Holland P.W.H. & Garcia-Fernöndez J. (1996) *Dev. Biol.* 173, 382–395.

Holm L. & Sander C. (1997) *Proteins* 28, 72–82.

Huang L., Hofer F., Martin G.S. & Kim S-H. (1998) *Nature Struct. Biol.* 5, 422–426.

Hunter T. (1987) *Cell* 49, 1–4.

Huynen M.A. & Bork P. (1998) *Proc. Natl. Acad. Sci. USA* **95**, 5849–5856.

Janin J. & Chothia C. (1985) *Meth. Enzymol.* **115**, 420–430.

Jung D., Yang B., Meyer J., Chamberlain J. S. & Campbell K. P. (1995) *J. Biol. Chem.* **270**, 27305–27310.

Kalhammer G., Bähler M., Schmitz F., Jöckel J. & Block C. (1997) *FEBS Lett.* **414**, 599–602.

Kimura M. (1983). *The Neutral Theory of Molecular Evolution*. Cambridge University Press, Cambridge.

Kistner U., Garner C.C. & Linial M. (1995) *FEBS Lett.* **359**, 159–163.

Koenig M., Monaco A.P. & Kunkel L.M. (1988) *Cell* **53**, 219–228.

Koonin E.V. & Abagyan R.A. (1997) *Nat. Genet.* **16**, 330–331.

Koonin E.V., Mushegian A.R. & Bork P. (1996) *Trends Genet.* **12**, 334–336.

Kumar S. & Hedges S.B. (1998) *Nature* **392**, 917–920.

Kuriyan J. & Cowburn D. (1997) *Annu. Rev. Biophys. Biomol. Struct.* **26**, 259–288.

Kurosky A., Barnett D.R., Lee T.H., Touchstone B., Hay R.E., Arnott M.S., Bowman B.H. & Fitch W.M. (1980) *Proc. Natl. Acad. Sci. USA* **77**, 3388–3392.

Lidov H.G.W. (1996) *Brain Pathol.* **6**, 63–77.

Lindqvist Y. & Schneider G. (1997) *Curr. Opin. Struct. Biol.* **7**, 422–427.

Mackay J. P. & Crossley M. (1998) *Trends Biochem.* **23**, 1–3.

Manley N.R. & Capecchi M.R. (1997) *Dev. Biol.* **192**, 274–288.

Martin J.L. (1995) *Structure* **15**, 245–250.

Miklos G.L.G. & Rubin G.M. (1996) *Cell* **86**, 521–529.

Murzin A.G. (1993) *Trends Biochem. Sci.* **18**, 403–405.

Mushegian A.R., Bassett D.E. Jr., Boguski M.S., Bork P. & Koonin E.V. (1997) *Proc. Natl. Acad. Sci. USA* **94**, 5831–5836.

Nadeau J.H. & Sankoff D. (1997) *Genetics* **147**, 1259–1266.

Nakamura T., Nishizawa T., Hagiya M., Seki T., Shimonishi M., Sugimura A., Tashiro, K. & Shimizu S. (1989) *Nature* **23**, 440–443.

Ohno S. (1970). *Evolution by Gene Duplication*. Springer Verlag, New York.

Owen M.C., Brenman S.O., Lewis J.H. & Carrell R.W. (1983) *N. Engl. J. Med.* **309**, 694–698.

Park I., Chung J., Walsh C.T., Yun Y. & Strominger J.L. (1995) *Proc. Natl. Acad. Sci. USA* **92**, 12338–12342.

Peters M.F., Adams M.E. & Froehner S.C. (1997) *J. Cell Biol.* **138**, 81–93.

Ponting C.P. (1997) *Protein Sci.* **6**, 459–463.

Ponting C.P. & Benjamin D.R. (1996) *Trends Biochem. Sci.* **21**, 422–425.

Ponting C.P. & Birney E. (1998) In *Protein structure prediction: Methods and Protocols*, Webster D.M. (ed.), Humana Press Inc., in press.

Ponting C. P., Blake D. J., Davies K. E., Kendrick-Jones J. & Winder S. J. (1996) *Trends Biochem. Sci.* **21**, 11–13.

Ponting C.P., Cai Y-D. & Bork P. (1997). *J. Mol. Med.* **75**, 467–469.

Radovick S., Nations M., Dy Y., Berg L.A., Weintraub B.D. & Wondisford F.E. (1992) *Science* **257**, 1115–1118.

Rafael J.A., Cox G.A., Corrado K., Jung D., Campbell K.P. & Chamberlain J.S. (1996) *J. Cell Biol.* **134**, 93–102.

Roberts R.G. & Bobrow M. (1998) *Hum. Mol. Genet.* **7**, 589–595.

Russell R.B. (1994) *Prot. Eng.* **7**, 1407–1410.

Russell R.B. & Ponting C.P. (1998) *Curr. Opin. Struct. Biol.* **8**, 364–371.

Sadoulet-Puccio H.M., Rajala M. & L.M. Kunkel (1997) *Proc. Natl. Acad. Sci. USA* **94**, 12413–12418.

Shibatohge M., Kariya K-I., Liao Y., Hu C-D., Watari Y., Goshima M., Shima F. & Kataoka T. (1998) *J. Biol. Chem.* **273**, 6218–6222.

Sicheri F., Moarefi I. & Kuriyan J. (1997) *Nature* **385**, 602–609.

Soriano P., Montgomery C., Geske R. & Bradley A. (1991) *Cell* **64**, 693–702.

Spring J. (1997) *FEBS Lett.* **400**, 2–8.

Stein P.L., Vogel H. & Soriano P. (1994) *Genes Dev.* **8**, 1999–2007.

Teglund S., McKay C., Schuetz E., van Deursen J.M., Stravopodis D., Wang D., Brown M., Bodner S., Grosveld G. & Ihle J.N. (1998) *Cell* **93**, 841–850.

Tinsley J.M., Blake D.J., Roche A., Fairbrother U., Riss J., Byth B.C., Knight A.E., Kendrick-Jones J., Suthers G.K., Love D.R., Edwards Y.H. & Davies K.E. (1992) *Nature* **360**, 591–593.

Vavvas D., Li X., Avruch J., Zhang X-F. (1998) *J. Biol. Chem.* **273**, 5439–5442.

Wilks A.F. (1989) *Proc. Natl. Acad. Sci. USA* **86**, 1603–1607.

Williams J.C., Wierenga R.K. & Saraste M. (1998) *Trends Biochem. Sci.* **23**, 179–187.

Winder S.J., Gibson T.J. & Kendrick-Jones J. (1995) *FEBS Lett.* **369**, 27–33.

Wishart M.J., Denu J.M., Williams J.A., Dixon J.E. (1995) *J. Biol. Chem.* **270**, 26782–26785.

Wright H.T. (1984) *J. Biol. Chem.* **259**, 14335–14336.

Xu W., Harrison S.C. & Eck M.J. (1997) *Nature* **385**, 595–602.

# 12 Structural Databases

## David Jones

Department of Biological Sciences, University of Warwick, Coventry, UK

## 12.1 INTRODUCTION

This chapter is intended as a brief overview of the role of structural databases in biology. At the very outset it is worth acknowledging the difference between a data bank and a database. Both terms refer to a computer-readable collection of data, but whereas a database implies that the data is organized into tables and accessed by means of a query language, a data bank implies very little beyond the fact that the data is stored in the form of computer files. In this chapter the two terms will be used interchangeably as it is now the case that although many publicly available collections of biological data are distributed in the form of simple flat files (and so best described as data banks) the data are generally maintained at the archival site in the form of a relational database.

Biological databases can be divided into two main types: primary databases and secondary databases. Primary databases are essentially pure archives of experimental data, whereas secondary databases are derived from the primary data, and contain 'value added information', e.g. open reading frame translations, motif definitions and crosslinks to other database entries. Until quite recently only primary databases existed for databases of structural information. The reason for this was probably the paucity of data, and the apparent lack of anything that might really 'add value' to the primary data. As will be discussed later, this view changed somewhat when it was realized that proteins with no apparent similarity in their sequences could prove to have very similar folds. Given that three-dimensional structure was evidently much more highly conserved than sequence, by tracking these similarities very distant evolutionary relationships between protein could be inferred. Another aspect of structural biology which no doubt contributed to the arrival of secondary structural databases was the greatly increased rate at which structures were being solved. This was a result of new experimental techniques (nuclear magnetic resonance (NMR) spectroscopy in particular) and significant automation of the long-serving X-ray crystallography itself. As the principal primary structural data-

*Genetics Databases*
ISBN 0-12-101625-0

base, the Protein Databank (PDB) (Bernstein *et al.*, 1977), became flooded with new data it became very apparent that the data urgently needed to be analysed and organized rather than simply stored.

## 12.2 THE BROOKHAVEN PROTEIN DATA BANK

The Protein Data Bank (PDB) currently maintained at Brookhaven (Bernstein *et al.*, 1977) is the world repository of experimentally determined protein structures (and also some other macromolecular structures such as nucleic acids and polysaccharides). Most of the entries are based on crystallographic studies, although an increasing number of NMR files are now included. The protein structures are stored in the form of flat text files in which the Cartesian coordinates of every atom in the protein are listed.

Despite its importance, PDB is not the only repository of biological structural information. Three other primary databanks exist, namely the Cambridge Structural Database (CSD), BioMagResBank (BMRB) and the Nucleic Acid Database (NDB). Although these three specialist databases are only of peripheral relevance to this chapter, it is nonetheless worth briefly describing them.

## 12.3 THE CAMBRIDGE STRUCTURAL DATABASE

The Cambridge Structural Database (CSD) (Watson, 1996) is the oldest of the three databases, and contains the three-dimensional structures of over 180 000 organic molecules (including small peptides), which have all been studied by either X-ray or neutron diffraction techniques. CSD is not often used for modelling the fold of a protein, but is very useful for modelling ligand-binding sites, and protein design. CSD comprises not just a set of data files, but a set of integrated software tools for querying and displaying the archived data.

For entry in the CSD the stored data can be divided into three categories: bibliographic information, molecular connectivity and three-dimensional coordinate data. The bibliographic data fields incorporate all of the bibliographic material for the entry and summarize the structural and experimental information for the crystal structure. The text and numerical information includes the authors' names and the full journal reference, as well as the crystallographic cell dimensions and the space group. The molecular connectivity fields essentially describe a conventional chemical diagram of the molecule, roughly corresponding to a full structural formula. A complete description of the atomic charges and bond types is stored in the form of a chemical connection table. It is worth noting that PDB does not include this connectivity information for the small molecule ligands included in many of its entries, and so CSD is often used to provide this missing information during modelling studies. As with PDB, the three-dimensional information in CSD provides the coordinates for each atom in the structure.

## 12.4 BIOMAGRESBANK

The BioMagResBank (BMRB: URL http://www.bmrb.wisc.edu) is the publicly-accessible depository for the results of NMR experiments on peptides, proteins, and nucleic acids, and is recognized by the International Society of Magnetic Resonance and by the IUPAC–IUBMB–IUPAB Inter-Union Task Group on the Standardization of Data Bases of Protein and Nucleic Acid Structures Determined by NMR Spectroscopy. Although BMRB overlaps with PDB to some extent, it aims to become the collection site for structural NMR data in proteins and nucleic acids. One aspect of BMRB which greatly separates it from PDB is that it also attempts to archive NMR-specific data such as chemical shifts, J-couplings and relaxation rates.

## 12.5 THE NUCLEIC ACID DATABASE

The Nucleic Acid Database (NDB) (Berman *et al.*, 1992) is simply the nucleic acid equivalent of PDB; these two data resources share common file formats and data is regularly exchanged between them. Again, there is some overlap with PDB since it also contains some structural information on nucleic acids, particularly when in complexes with protein structures.

## 12.6 A TYPICAL PDB ENTRY

Figure 12.1 shows an excerpt from a typical PDB entry. A PDB file consists of two main sections, one containing information records and one containing the experimental data itself. Despite the apparent simplicity and readability of the PDB file format, a number of important pieces of data cannot be readily extracted from the file by a computer program. Often key pieces of experimental data are included in the 'REMARK' records, which do not conform to any particular formatting guidelines.

Despite the fact that a PDB file is designed to be read by a human, reading the information is not the same as understanding it. Obviously, a molecular graphics program can be used to display the three-dimensional coordinates of the structure, but this does not allow for easy browsing of PDB. One solution to this problem is to create simple summaries of each PDB entry so that important features of a particular entry can be seen at a glance. One popular collection of such summaries is maintained at University College London, and is called PDBSUM (Laskowski *et al.*, 1997). Figure 12.2 shows the same PDB file (as shown in Fig. 12.1) displayed as a simple graphical summary. PDBSUM includes not only information extracted from the PDB entry itself, but also

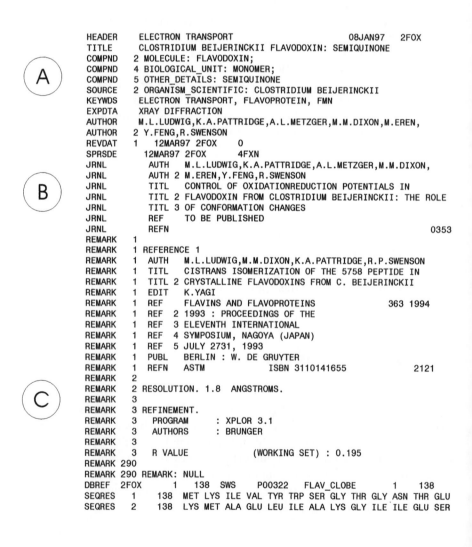

```
      HEADER    ELECTRON TRANSPORT                      08JAN97   2FOX
      TITLE     CLOSTRIDIUM BEIJERINCKII FLAVODOXIN: SEMIQUINONE
      COMPND    2 MOLECULE: FLAVODOXIN;
      COMPND    4 BIOLOGICAL_UNIT: MONOMER;
 A    COMPND    5 OTHER_DETAILS: SEMIQUINONE
      SOURCE    2 ORGANISM_SCIENTIFIC: CLOSTRIDIUM BEIJERINCKII
      KEYWDS    ELECTRON TRANSPORT, FLAVOPROTEIN, FMN
      EXPDTA    XRAY DIFFRACTION
      AUTHOR    M.L.LUDWIG,K.A.PATTRIDGE,A.L.METZGER,M.M.DIXON,M.EREN,
      AUTHOR    2 Y.FENG,R.SWENSON
      REVDAT    1    12MAR97 2FOX    0
      SPRSDE       12MAR97 2FOX         4FXN
      JRNL          AUTH   M.L.LUDWIG,K.A.PATTRIDGE,A.L.METZGER,M.M.DIXON,
      JRNL          AUTH 2 M.EREN,Y.FENG,R.SWENSON
 B    JRNL          TITL   CONTROL OF OXIDATIONREDUCTION POTENTIALS IN
      JRNL          TITL 2 FLAVODOXIN FROM CLOSTRIDIUM BEIJERINCKII: THE ROLE
      JRNL          TITL 3 OF CONFORMATION CHANGES
      JRNL          REF    TO BE PUBLISHED
      JRNL          REFN                                              0353
      REMARK    1
      REMARK    1 REFERENCE 1
      REMARK    1 AUTH   M.L.LUDWIG,M.M.DIXON,K.A.PATTRIDGE,R.P.SWENSON
      REMARK    1 TITL   CISTRANS ISOMERIZATION OF THE 5758 PEPTIDE IN
      REMARK    1 TITL 2 CRYSTALLINE FLAVODOXINS FROM C. BEIJERINCKII
      REMARK    1 EDIT   K.YAGI
      REMARK    1 REF    FLAVINS AND FLAVOPROTEINS            363 1994
      REMARK    1 REF  2 1993 : PROCEEDINGS OF THE
      REMARK    1 REF  3 ELEVENTH INTERNATIONAL
      REMARK    1 REF  4 SYMPOSIUM, NAGOYA (JAPAN)
      REMARK    1 REF  5 JULY 2731, 1993
      REMARK    1 PUBL   BERLIN : W. DE GRUYTER
      REMARK    1 REFN   ASTM            ISBN 3110141655            2121
      REMARK    2
 C    REMARK    2 RESOLUTION. 1.8   ANGSTROMS.
      REMARK    3
      REMARK    3 REFINEMENT.
      REMARK    3 PROGRAM    : XPLOR 3.1
      REMARK    3 AUTHORS    : BRUNGER
      REMARK    3
      REMARK    3 R VALUE         (WORKING SET) : 0.195
      REMARK 290
      REMARK 290 REMARK: NULL
      DBREF 2FOX    1   138  SWS    P00322    FLAV_CLOBE      1    138
      SEQRES    1   138 MET LYS ILE VAL TYR TRP SER GLY THR GLY ASN THR GLU
      SEQRES    2   138 LYS MET ALA GLU LEU ILE ALA LYS GLY ILE ILE GLU SER
```

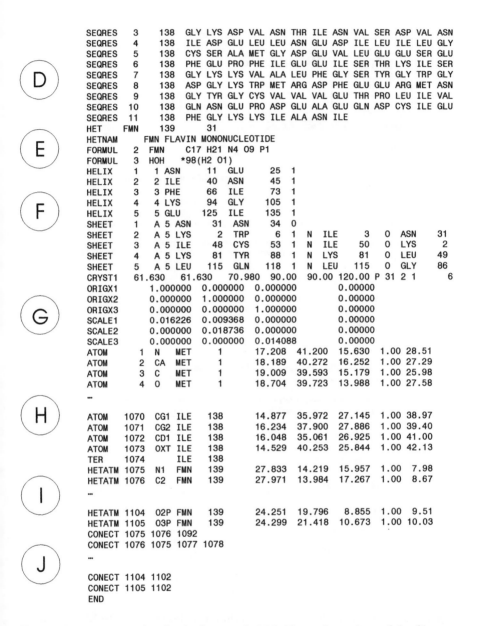

```
SEQRES    3   138   GLY LYS ASP VAL ASN THR ILE ASN VAL SER ASP VAL ASN
SEQRES    4   138   ILE ASP GLU LEU LEU ASN GLU ASP ILE LEU ILE LEU GLY
SEQRES    5   138   CYS SER ALA MET GLY ASP GLU VAL LEU GLU GLU SER GLU
SEQRES    6   138   PHE GLU PRO PHE ILE GLU GLU ILE SER THR LYS ILE SER
SEQRES    7   138   GLY LYS LYS VAL ALA LEU PHE GLY SER TYR GLY TRP GLY
SEQRES    8   138   ASP GLY LYS TRP MET ARG ASP PHE GLU GLU ARG MET ASN
SEQRES    9   138   GLY TYR GLY CYS VAL VAL VAL GLU THR PRO LEU ILE VAL
SEQRES   10   138   GLN ASN GLU PRO ASP GLU ALA GLU GLN ASP CYS ILE GLU
SEQRES   11   138   PHE GLY LYS LYS ILE ALA ASN ILE
HET     FMN   139        31
HETNAM        FMN FLAVIN MONONUCLEOTIDE
FORMUL    2   FMN     C17 H21 N4 O9 P1
FORMUL    3   HOH    *98(H2 O1)
HELIX     1   1 ASN     11 GLU     25  1
HELIX     2   2 ILE     40 ASN     45  1
HELIX     3   3 PHE     66 ILE     73  1
HELIX     4   4 LYS     94 GLY    105  1
HELIX     5   5 GLU    125 ILE    135  1
SHEET     1   A 5 ASN   31 ASN     34  0
SHEET     2   A 5 LYS    2 TRP      6  1  N ILE      3   O ASN     31
SHEET     3   A 5 ILE   48 CYS     53  1  N ILE     50   O LYS      2
SHEET     4   A 5 LYS   81 TYR     88  1  N LYS     81   O LEU     49
SHEET     5   A 5 LEU  115 GLN    118  1  N LEU    115   O GLY     86
CRYST1   61.630    61.630    70.980  90.00   90.00 120.00 P 31 2 1       6
ORIGX1        1.000000  0.000000  0.000000        0.00000
ORIGX2        0.000000  1.000000  0.000000        0.00000
ORIGX3        0.000000  0.000000  1.000000        0.00000
SCALE1        0.016226  0.009368  0.000000        0.00000
SCALE2        0.000000  0.018736  0.000000        0.00000
SCALE3        0.000000  0.000000  0.014088        0.00000
ATOM      1   N    MET     1      17.208  41.200  15.630  1.00 28.51
ATOM      2   CA   MET     1      18.189  40.272  16.252  1.00 27.29
ATOM      3   C    MET     1      19.009  39.593  15.179  1.00 25.98
ATOM      4   O    MET     1      18.704  39.723  13.988  1.00 27.58
...
ATOM   1070   CG1  ILE   138      14.877  35.972  27.145  1.00 38.97
ATOM   1071   CG2  ILE   138      16.234  37.900  27.886  1.00 39.40
ATOM   1072   CD1  ILE   138      16.048  35.061  26.925  1.00 41.00
ATOM   1073   OXT  ILE   138      14.529  40.253  25.844  1.00 42.13
TER    1074        ILE   138
HETATM 1075   N1   FMN   139      27.833  14.219  15.957  1.00  7.98
HETATM 1076   C2   FMN   139      27.971  13.984  17.267  1.00  8.67
...
HETATM 1104   O2P  FMN   139      24.251  19.796   8.855  1.00  9.51
HETATM 1105   O3P  FMN   139      24.299  21.418  10.673  1.00 10.03
CONECT 1075 1076 1092
CONECT 1076 1075 1077 1078
...
CONECT 1104 1102
CONECT 1105 1102
END
```

**Figure 12.1** An example of a single entry in PDB. The main sections of the file are marked as follows: (A) header records and description of protein, (B) author and bibliographic records, (C) crystallographic quality parameters (resolution and R-factor among others), (D) amino acid sequence, (E) molecular formulae of protein and ligands, (F) description of secondary structural elements in structure, (G) crystal unit cell parameters, (H) coordinates of protein atoms, (I) coordinates of ligand and co-factor atoms and (J) bond connectivity for ligands and co-factors.

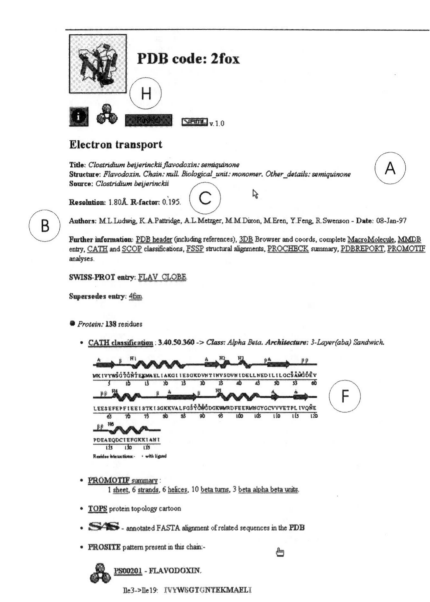

**Figure 12.2** An example of the same PDB entry shown in Figure 12.1, but this time displayed in the form of a graphical summary. A subset of the information found in the raw data file is presented: (A) header records describing the biological context of the protein molecule, (B) author records, (C) crystallographic quality values (resolution and R-factor), (D) amino acid sequence, (E) not represented, (F) summary of secondary structure, (G) not represented, (H–I) atomic coordinates (rendered graphically) and (J) not represented.

links to other secondary databases (see later). Apart from a small cartoon representation of the three-dimensional structure itself, including any bound ligands, PDBSUM also shows a simplified representation of the protein's secondary structure, along with any structural motifs identified.

## 12.7 PROTEIN STRUCTURE CLASSIFICATION RESOURCES

Like all public bioinformatics resources, the PDB has grown dramatically over the last 20 years and there are now over 6000 structures and over 100 new structures added to the data bank each month. This wealth of structural data can pose a problem to the biologists using PDB as a source of information about a single protein or a family of related structures. To get maximum biological information from PDB, all the members of a protein family should ideally be identified and given that many families are large, this is not a trivial task. For example, there are now more than 120 entries for chymotrypsin-like serine proteinases in PDB. How does a non-expert access these structures in order to compare and contrast them? A further problem with PDB is that for casual use of the databank, much of the information in the databank is 'redundant'. Nearly two-thirds of the structures in PDB have nearly identical sequences to one or more other proteins, and so will not be expected to contribute much additional biological data.

To help make sense of the structures stored in PDB, several groups around the world have started projects aimed at classifying every structure in PDB into helpful groupings.

In the case of PDB entries which have very similar protein sequences, it goes without saying that all these structures will be essentially the same. However, at the other end of the scale, proteins with no statistically significant sequence similarity may also adopt similar structures and in some cases the similarity is so striking that an evolutionary relationship may be inferred. Since these similarities can only be revealed by comparing the three-dimensional structures, a comprehensive classification of PDB structures allows rapid examination of these relationships.

In recent years, the number of structural similarities which have been observed in the absence of sequence similarity has grown rapidly. Given a newly solved protein structure with no obvious sequence similarity to a protein of known three-dimensional structure it is useful to ask what the chance is that this new protein has a novel protein fold. The answer (at the time of writing) is surprisingly low: 30–40%. Thus, there are already a large number of such fold similarities to be identified. However, structure comparison and classification is useful not only for evaluating distant evolutionary relationships, it is also useful for providing data for structure prediction methods, and in particular the 'fold recognition methods'. These methods (see later for more details), are able to assess the compatibility of a target protein sequence with a library of structures, using a statistically derived energy function.

## 12.8   METHODS FOR COMPARING PROTEIN STRUCTURES

Clearly, before the structures in PDB can be classified some means is needed to identify similarities between protein folds. The simplest way is of course to make the comparisons by eye and this can be a very successful approach. However, classifying thousands of protein structures by eye is not really practical and creates the problem of subjectivity in constructing the classification. Consequently, algorithms which automatically identify structural relationships between proteins have been actively investigated in recent years. A number of different computer science algorithms have been exploited including dynamic programming (Taylor & Orengo, 1989), least-squares superposition (Pascarella & Argos, 1992; Diamond, 1992), simulated annealing (Sali & Blundell, 1990), graph theory (Grindley *et al.*, 1993), distance matrix comparison (Holm & Sander, 1993) and geometric hashing (Nussinov & Wolfson, 1991). Methods for structure comparison have been recently reviewed by Brown *et al. (1996)*.

## 12.9   AVAILABLE CLASSIFICATION SCHEMES

Preliminary attempts at providing a classification of known protein structures were published in papers in the form of large tables (e.g. Orengo *et al.*, 1993). However, these data quickly became out of date. Modern attempts have taken full advantage of electronic Web-based publication and these classifications are updated at regular intervals. With colleagues at University College London, UK, and the Helix Research Institute in Japan I have been involved with developing a hierarchical classification of protein domain structures, called CATH (Orengo *et al.*, 1997). The CATH classification (see Section 12.10) is, of course, not the only classification scheme available, and, indeed, was not the first.

## 12.9.1   SCOP

The very first, and best known protein structure classification scheme is SCOP (Murzin *et al.*, 1995), a classification which, like CATH, also attempts to describe the similarities between structures in a hierarchical manner. The main difference from CATH is that SCOP is essentially constructed manually. Even more surprising is that although the overall maintenance of SCOP represents a team effort, the classification itself is essentially carried out by a single individual (Dr Alexey Murzin). Despite the difficulty in maintaining such a complex classification manually, SCOP remains up to date and still represents the most extensive description of evolutionary relationships between protein structures in PDB. Despite the excellence of the biological information in SCOP, it is somewhat weaker in the way it classifies protein structures beyond obvious evolutionary relationships. For example, the remarkable similarity between the fold of colicin A and the globin family in the absence of any obvious common ancestry is not properly represented in SCOP. Although the similarity is remarked upon in a short comment, it is not represented in the classification scheme itself.

## 12.9.2　Nonhierarchical schemes

The other two main classification resources, DALI (Holm & Sander, 1997) and VAST (Hogue *et al.*, 1996), are not hierarchical classifications but are really just methods which have been used to make large-scale structure comparisons. As a result, both DALI and VAST provide the scores for each pairwise structure comparison and can display sequence alignments of those structures which are closest to the structure of interest. In the DALI method this is provided by the FSSP (Holm & Sander, 1997) database whereas the VAST approach generates a database called MMDB. Although neither scheme provides the same overview of protein structure space that is available by the hierarchical schemes, DALI and VAST are particularly useful for further computer processing of structural similarities detected. For example, DALI allows a researcher to query the database to find out whether a newly determined structure is similar to any of those already deposited in PDB. Conversely, VAST uses the structural similarities detected to make links between entries in the Medline database at the NCBI, along with links to the sequence databanks. Use of the VAST database enables complex searches on both sequences and structures to be made very rapidly across the Web.

　　Other classification schemes which are also in common use are 3Dee (http://circinus.ebi.ac.uk:8080/3Dee/help/help_intro.html) dbase (Sowdhamini *et al.*, 1996) and 3D_ali (Pascarella & Argos, 1992).

## 12.10　CONSTRUCTING THE CATH CLASSIFICATION

CATH stands for Class, Architecture, Topology and Homology and describes a hierarchy of four levels of increasingly detailed structural properties for each protein in the PDB. The main idea being that protein domains which share the same CATH classification code will also have structures which are globally similar. Figure 12.3 shows a simple example of these four CATH levels. Further levels are also defined based on the degree of redundancy at the sequence level. For example, an S-level is defined such that all proteins with >35% sequence identity are assigned identical CATH codes.

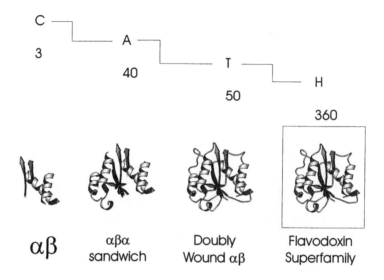

**Figure 12.3** Example of a single entry in the CATH protein structure classification database is shown, which demonstrates how an entry in CATH is classified. Again, the entry corresponding to PDB entry 2FOX is shown.

## 12.10.1 Definition of structural class

The C-level is essentially defined as the composition of secondary structure type and distribution of secondary structural elements along the polypeptide chain.

The four principal classes are as follows:

(1) $\alpha$: predominantly $\alpha$-helices, few $\beta$-strands
(2) $\beta$: predominantly $\beta$-strands, few $\alpha$-helices
(3) $\alpha/\beta$: any mixture of $\alpha/\beta$ structure
(4) Low SS: low percentage of residues in secondary structures

## 12.10.2 Definition of architecture

The A-level classifies protein folds on the basis of the gross geometric arrangement of secondary structural elements, but not taking into account the connectivity of these elements (i.e. the order in the sequence does not matter). For example, in the mainly-α class (C = 1), only three different architectures are defined where A = 10 indicates an orthogonal arrangement of helices, A = 20 indicates a helix bundle and A = 30 represents a mixture of the two types of helix packing. Architectural types for the all-β and α/β classes are much more varied.

## 12.10.3 Definition of topology

The T-level essentially is a subdivision of the A-level, in that not only is the geometric arrangement of secondary structures taken into account, but the chain connectivity is also compared. Protein domains with the same C, A and T numbers are said to have the same overall fold.

## 12.10.4 Definition of homology level

At this level of the classification some subjectivity arises. For the most similar structures, the question arises as to whether the structural similarity is a result of the two protein domains having evolved from a common ancestor. Where there is evident sequence similarity between domains, then common ancestry is almost certain. However, in the absence of sequence similarity, the decision is far harder to make. Essentially, two clues are considered when assigning protein domains to the same H-level:

- Are the structures 'unusually' similar?
- Do the protein domains share some common functionality?

When considering these two principal questions, a literature search is often also required to try to clarify the situation. Nevertheless, there remains no reliable means to decide whether or not two proteins truly share common ancestry – even the assignments in the SCOP database are based on a mixture of biological insight and biochemical knowledge.

## 12.10.5 An outline of the classification procedure

As already noted, proteins which have very similar protein sequences are expected to have similar folded conformations. In view of this, in constructing

CATH, as with other classification procedures sequence comparisons are initially carried out between all the proteins being classified. Since it is now well established that sequences with significant similarity (generally >30% sequence identity) almost certainly adopt the same fold, sequence alignments can be used to assign proteins to the same family (assigning at the H-level). In the CATH database, a cautious cut-off is used to identify related structures. At present, simple sequence comparison reduces the 6000 protein chains found in the PDB to just over 1500 clusters.

Although the clustering of protein chains into sequence families is straightforward, one complication lies in the choice of a single representative from each family to use in later stages of the classification. Two things need to be considered: the quality of the raw data, and whether the protein chain is a naturally occurring molecule or an artificially constructed mutant. CATH chooses structures using the following categories (in order of decreasing priority):

(1)  High-resolution X-ray determined structure of native protein
(2)  High-resolution X-ray determined structure of mutant protein
(3)  NMR structure of native protein
(4)  NMR structure of mutant protein
(5)  A-carbon only model
(6)  Crystal structures with resolution >3.0Å
(7)  Theoretical model structures

Structures which fall into categories 6 and 7 are not included in the CATH database.

Having reduced the redundancy in the data, and chosen a single 'best' representative for each family of sequences, the next step is to decompose each polypeptide chain into structural domains (compact substructures which look likely to be independent folding units). Although this procedure makes use of automatic domain assignment methods, in many cases some visual inspection is required to check that a reasonable set of domains has been produced for each representative structure.

As already noted, however, there are now many examples of proteins sharing no apparent sequence similarity, but which fold in a similar way; to detect this kind of similarity, structure comparison methods must be applied. However, since structure comparison is generally very slow, the number of necessary comparisons between all the representative domains is reduced by first assigning proteins to their respective structural classes (i.e. all-$\alpha$, all-$\beta$, mixed-$\alpha$/$\beta$), and then only comparing proteins in the same class.

To generate the CATH database of protein families, structural comparisons are performed on representatives from each of the sequence based families using the program SSAP (Taylor & Orengo, 1989).

The above methods account for the C and T (and mostly H) levels of the CATH scheme. For proteins which exhibit novel folds, the architecture (i.e.

arrangement of secondary structures in the protein fold without consideration of chain connectivity) must be determined manually based on established descriptions from the literature (e.g. β-sandwich or β-barrel). Clearly for folds which are not novel, the architectural assignment will already have been made.

## 12.10.6   Protein space as described by CATH

Having completed the classification it is possible to get a good overview of the current knowledge of protein structure space. Version 1.1 (the first publicly available version) of CATH contains 8078 domains which are derived from 5995 protein chains of the Sept 1996 Brookhaven PDB release. The number of entries in each level of the very latest release of CATH are summarized in Fig. 12.4. The redundancy of this data, based on sequence similarity, is immediately apparent from the fact that when only a single representative is taken from each S-level (where all members have less than 35% residue identity with each other), the 8078 protein domains reduce to just 1068 families.

By applying structure comparison and requiring that potential similarities also have a concomitant functional relationship these 1068 S-level families can be clustered into 645 H-levels.

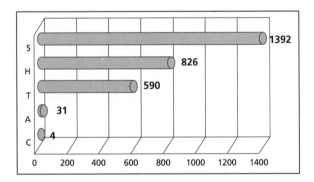

**Figure 12.4**   The current distribution of protein domains at five levels (C,A,T,H and the sequence level S) in the latest release of CATH (V1.4).

## 12.10.7   Class and architecture

At the C-level (Class) the 1068 sequence families distribute essentially as anticipated with 24.3% in α, 21.2% in β and 52.2% in α/β. Only 1.3% of the families are placed outside these three categories (i.e. in the unclassified category). As mentioned earlier, there are a noticeably different number of architectures per class, varying from two discernible architectures in the α class, nine in the α/β class and 17 in the β class. Why the α class only has two main architectures is explained simply by the fact that helices pack against each other in protein folds using an almost continuous distribution of angles, which result in a lack of distinct tertiary patterns when connectivity is ignored. In contrast, architectures, formed by β-strands are much more restricted (owing to hydrogen bonding constraints), and so have much more distinguishable shapes.

## 12.10.8    Topology and homology

CATH 1.1 incorporates 505 different domain topologies (T-level) and 645 homologous superfamilies (H-level). Interestingly, the vast majority of T-levels contain only a single H-level. In other words, most protein folds are unique to those proteins which are related by common ancestry. For example, only proteins in the lysozyme superfamily itself exhibit the lysozyme fold. However, in certain cases more than one H-level exists, i.e. a particular fold is not unique to a particular superfamily of proteins. These folds are known as 'superfolds', and 10 such folds have been observed to date: the TIM barrel, the doubly-wound α/β structure, the immunoglobulin fold, the jelly roll, four-helix bundles, the globin fold, the β trefoil (interleukin-1β fold), the UB roll, the α+β plait and the OB fold.

## 12.11    MAKING USE OF STRUCTURAL DATABASES

It is not possible, of course, to list every conceivable use of structural databases in biology. Rather than trying to list all the possible applications of structural databases, we will look at just one very common application, namely modelling the conformation of a newly characterized protein sequence.

## 12.11.1    I have this sequence . . .

The main rule to follow when attempting to predict the structure of a newly characterized protein sequence is to use the simplest and most reliable method that can provide the level of structural detail that is required. It is important to be realistic, however. Many prediction methods can only provide a crude model for your protein if they work at all. In these cases it may be possible to get some insight into the gross topology of the folded protein chain but it will not be possible to predict detailed features (such as the positions of amino acid side-chains). If your protein has a high degree of sequence similarity to a protein of known structure, then standard comparative modelling techniques can be applied. In this case, a fairly accurate model can be expected. If you are able to build a good comparative model of your protein, then do not waste time with more complicated and less reliable methods, such as secondary structure prediction, threading or *ab initio* folding simulations. In rare circumstances there might be some cause to enhance a model using more advanced methods, for example where an additional domain has been inserted into a structure, but in the great majority of cases the simplest approaches are best.

The basic question that needs to be addressed when analysing a new sequence is the question of which entry or entries in the structural database (PDB or one of the classification databases such as CATH) is likely to resemble the native structure of the new sequence most closely. To demonstrate how easy (or difficult) this can be, two examples will be shown from the Second Critical Assessment in Structure Prediction experiment (CASP2) (Levitt, 1997), which is an international collaboration between theoreticians and experimental workers to determine the current level of success that can be expected from protein structure prediction and modelling. From these experiments it has become clear that it is important to distinguish between cases where the target protein shows distant homology to a protein of known three-dimensional structure, and cases where no common ancestry is apparent. This suggests that an attempt to model a newly characterized sequence should be divided into two steps: one which concentrates on identifying a distant evolutionary relationship and, if necessary, a further step which attempts to identify only a common fold between two proteins (fold recognition).

## 12.11.2 Step 1: searching for similar sequences

Before moving onto more advanced methods for prediction it is advisable to expend a great deal of effort in detecting homology between the target protein and proteins of known three-dimensional structure. This seems like an obvious thing to do, but it is commonly seen that workers spend very little time on this step and end up leading themselves astray, or falsely concluding that their protein does not resemble any known three-dimensional structure.

There are a number of sequence comparison software packages which can be obtained freely from the authors. At the time of writing, one package in particular is highly recommended: PSI-BLAST (Altschul *et al.*, 1997). PSI-BLAST is the latest incarnation of the BLAST (Altschul *et al.*, 1990) sequence comparison program, and, amongst other things, is now capable of generating gapped alignments. Of even more interest are the new Position Specific Iteration (PSI) features of PSI-BLAST. Here, rather than searching a data bank of sequences with a single sequence and then finishing, the program builds a profile (i.e. a multiple sequence alignment) based on the initial search results and uses this profile to search the data bank again. If the profile pulls any more sequences out of the data bank, then these new sequences are added to the profile and the new profile used again to search for more sequences. This procedure can be terminated after a fixed number of iterations, or can be allowed to continue until no more sequences are detected (according to pre-set convergence criteria).

To make use of PSI-BLAST for detecting a relationship between a target protein sequence and an entry in a structure database, it is necessary to create a data bank of sequences for which the three-dimensional structure is known. In other words, it is necessary to extract sequence information from Brookhaven PDB formatted files (Bernstein et al., 1977) and convert these data into a flat-file sequence format. Fortunately, converted sequence files are readily available by WWW or ftp (e.g. ftp://ftp.ebi.ac.uk/pub/databases/pdb_seq).

As a good example of the use of PSI-BLAST in assigning known three-dimensional structures to novel sequences, we will take as an example exfoliative toxin A from *Staphylococcus aureus*. To use PSI-BLAST to maximum effect, a hybrid sequence data bank is required, which contains the sequences of all proteins currently archived in the Brookhaven PDB along with other protein sequences taken from a standard protein sequence data bank such as OWL (Bleasby et al., 1994). In this example, 3576 nonredundant sequences were extracted from PDB and added to 244 827 sequences taken from OWL. Why add these 244 827 additional sequences? Recall that PSI-BLAST works by constructing sequence profiles as it runs. Sequence profiles are generally improved by including as many diverse sequences as possible and so these additional sequences can greatly extend the range of PSI-BLAST in detecting homology between the target sequence and a PDB sequence.

Here, PSI-BLAST is run on this sequence using the parameters -h 0.001 -j 10. The first iteration from PSI-BLAST produces the matches shown in Fig. 12.5a. These hits are all significant, but at this stage none of the hits are to any of the sequences taken from PDB. However, on the third PSI-BLAST iteration, the output shown in Fig. 12.5b is produced. Although quite a long way down the list of hits, a statistically significant match is reported between the target protein and PDB entry 1TRY (E-value = $7 \times 10^{-6}$). In subsequent iterations the E-value for this match drops to as little as $10^{-20}$, which is clearly significant. Figure 12.6 shows the crystal structures of both the target protein (Vath et al., 1997) and the matched protein (Rypniewski et al., 1995) as found in PDB. It is quite clear from the Molscript (Kraulis, 1991) diagrams that the proteins share a common fold and, owing to the obvious functional and sequence similarity, that the proteins share common ancestry.

In the event of a similarity being detected at this stage, then it is advisable to stop here and consider the construction of a homology model. However, this is not as easy as it sounds. Unless the similarity is very obvious, it will not be easy to produce a correct alignment between the target protein and the protein of known structure. However, given the difficulties inherent in using more involved prediction methods, the effort is likely to be worthwhile. PSI-BLAST itself can output an alignment which can be used for modelling, but often these alignments can be improved by the use of more sophisticated multiple sequence alignment methods or manual adjustment.

**(a)**

```
ETA_STAAU EXFOLIATIVE TOXIN A PRECURSOR (EC 3.4.21.-) (EPIDERMO...    499  e-141
ETB_STAAU EXFOLIATIVE TOXIN B PRECURSOR (EC 3.4.21.-) (EPIDERMO...    197  4e-50
S21758 glutamic acid-specific endopeptidase - Staphylococcus au...     94  5e-19
STSP_STAAU GLUTAMYL ENDOPEPTIDASE PRECURSOR (EC 3.4.21.19) (STA...     93  1e-18
SAU60589 SAU60589 NID: g1407783 - Staphylococcus aureus.              84  7e-16
SAU63529 SAU63529 NID: g1488694 - Staphylococcus aureus.              72  2e-12
S25140 serine proteinase homolog - Enterococcus faecalis             65  3e-10
C64647 serine proteinase (EC 3.4.21.-) - Helicobacter pylori (s...     50  1e-05
D78376 D78376 NID: g1526427 - Yersinia enterocolitica (strain:W...     48  4e-05
YEHTRA YEHTRA NID: g1419350 - Yersinia enterocolitica.               48  4e-05
CJHTRA CJHTRA NID: g2077988 - Campylobacter jejuni.                  46  1e-04
CJU27271 CJU27271 NID: g881374 - Campylobacter jejuni.               46  1e-04
E1181491 YKDA.                                                       45  5e-04
DEGQ_ECOLI PROTEASE DEGQ PRECURSOR (EC 3.4.21.-).                    43  0.001
```

**(b)**

```
..
POLG_PPVNA GENOME POLYPROTEIN (CONTAINS: N-TERMINAL PROTEIN; HE...    53  1e-06
PSU05771 PSU05771 NID: g1335723 - Peanut stripe virus.               53  1e-06
PSU34972 PSU34972 NID: g1016234 - Peanut stripe virus.               53  1e-06
YNM3_YEAST HYPOTHETICAL 110.9 KD PROTEIN IN SPC98-TOM70 INTERGE...    53  1e-06
PVCHYMOA PVCHYMOA NID: g2462646 - Penaeus vannamei.                  52  3e-06
PVCHYMOB PVCHYMOB NID: g2462648 - Penaeus vannamei.                  51  4e-06
pdb|1TRY|1TRY  trypsin                                               51  7e-06 -
CTR2_VESOR CHYMOTRYPSIN II (EC 3.4.21.1).                            49  2e-05
STMSAMP20 STMSAMP20 NID: g474021 - Streptomyces albogriseolus (...     49  2e-05
BCU19287 BCU19287 NID: g625062 - Bean common mosaic virus.           49  2e-05
YMU425966 YMU42596 NID: g1552411 - Yam mosaic virus.                 49  2e-05
.
.
```

**Figure 12.5** (a) First pass results generated by PSI-BLAST for target sequence. (b) Section of results from the third pass of the PSI-BLAST algorithm.

Trypsin

Exf-A

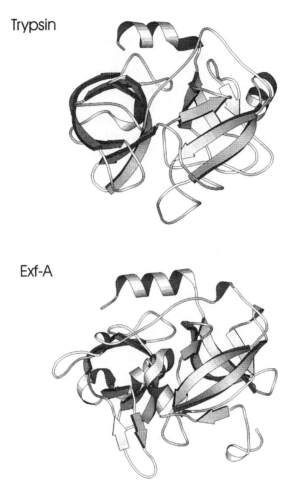

**Figure 12.6**   Molscript diagram showing the similarity between target protein and the matched structure for the search in Fig. 12.5b.

## 12.11.3    Step 2: fold recognition

The term 'fold recognition' generally refers to methods which take a single sequence and attempt to find a structure (from a library of known structures) which is most likely to resemble the native conformation of the sequence. Many methods for this have been proposed over the last few years, but in general terms a measure of sequence-to-structure compatibility is used, along with some means for performing the alignment (or fitting) of the sequence onto each candidate structure. These measures are either based on one-dimensional sequence properties, such as solvent accessibility (Bowie *et al.*, 1991; Alexandrov *et al.*, 1995; Fischer & Eisenberg, 1996; Russell *et al.*, 1996), or incorporate pseudo-energetic potentials based on inter-residue pairwise interactions (Hendlich *et al.*, 1990; Jones *et al.*, 1992; Maiorov & Crippen, 1992; Bryant & Lawrence, 1993; Flöckner *et al.*, 1995). The substitution matrices described by Overington *et al.* (1992) represent an alternative approach to measuring sequence-structure compatibility for inverse protein folding.

## 12.12   HOW DOES THREADING WORK?

Although compatibility between a sequence and a structure is usually measured by calculating energy terms for pairwise interactions between amino acids, these specific interactions are often not conserved between analogous proteins (Russell & Barton, 1994) and yet threading methods do seem to work (Levitt, 1997; Jones *et al.*, 1995). The first point to make in perhaps explaining this is that the accessibility of a residue is not well conserved across families of proteins (Flores *et al.*, 1993). Relative insertions and deletions and even mutations between two related proteins can greatly alter the accessibility of particular equivalent sites in the two structures. Profile-based fold recognition methods (e.g. three-/one-dimensional profiles), however, make the assumption that the pattern of accessibility between two divergent protein structures is perfectly conserved, and it is this assumption that results in their relatively poor performance. Threading methods, however, model the environment of a residue by summing the surrounding hydrophobic pair interactions. These pair interaction environments change as the threading alignment changes and it is this sensitivity of residue environments to changes in the sequence–structure alignment that results in the increased predictive power of threading methods. Although this explains why threading works even when specific contacts are not conserved, it also explains why sequence–structure alignments are generally of poor quality when compared with structure–structure alignments. The explanation being that there are typically many ways of arranging a sequence on a structural template such that its hydrophobic residues are adequately screened by other residues. If the potentials encoded more specific information, then it might be hoped that the correct alignment might be better distinguished from the incorrect alternatives, but despite a lot of effort in a number of different labs, there has been little improvement in the quality of potentials used for fold recognition.

## 12.12.1    Strong and weak fold recognition

At this point it is worth considering two versions of the protein fold recognition problem: weak fold recognition and strong fold recognition. In the strong form of the problem we seek a set of potentials (and a method for performing the sequence structure alignment) which will reliably recognize the closest matching fold for a given sequence from the thousands of alternatives (we have estimated as many as 7000 naturally occurring folds; Orengo *et al.*, 1994). This is the stated goal of work published to date. Unfortunately, this form of the problem may well prove to be too difficult to solve. Why should the strong fold recognition problem be insoluble? Quite simply, because the real protein free energy function is itself almost certainly incapable of satisfying this requirement. Current experimental evidence on protein folding pathways suggests that proteins do not explore all possible topologies, but a very limited set restricted by local restraints imposed early in the folding process.

The weak fold recognition problem is a far more realistic formulation of the problem. Here the problem is to recognize and exclude folds which are not compatible with the given sequence, with the aim of arriving at a short-list of possible conformations for the protein being modelled. By applying other filters such as similarity in chain length, similarity of function or perhaps experimentally derived constraints, it might be hoped that the list could be whittled down to just a few possibilities (ideally just one). Despite the initial aim being to solve the strong recognition problem, there is little evidence that any of the currently known fold recognition methods have gone further than weak recognition. However, the success of fold recognition methods in predicting protein tertiary structure when subjected to rigorous blind testing (Jones *et al.*, 1995; Lemer *et al.*, 1995; Levitt, 1997) suggests that intelligent use of weak fold recognition may nevertheless be quite powerful.

The following text outlines the practical application of a widely-used threading program, THREADER (Jones *et al.*, 1992, 1995). This is not the only threading method available (see earlier), but is one of the few programs to be made widely available to the academic community. Although the specifics of this section relate solely to THREADER, the principles can be applied to almost any threading program.

As an example, consider the following sequence (taken from the THREADER manual):

```
>CBDN CBDN1 from Cellulomonas fimi, 152a.a.
ASPIGEGTFDDGPEGWVAYGTDGPLDTSTGALCVAVPAGSAQYGVGVVLN
GVAIEEGTTYTLRYTATASTDVTVRALVGQNGAPYGTVLDTSPALTSEPR
QVTETFTASATYPATPAADDPEGQIAFQLGGFSADAWTLCLDDVALDSEV
EL
```

This sequence has no apparent sequence similarity to any other sequence with a known structure, even when tested using PSI-BLAST.

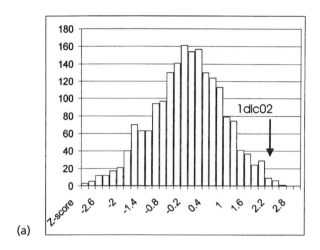

(a)

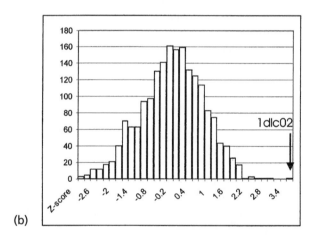

(b)

**Figure 12.7** (a) The raw threading results for the cellulose-binding domain from *Cellulomonas fimi*, again one of the CASP2 targets (Levitt, 1997). Although the top 10 folds includes one fold protein (1dlc02 – domain 2 of PDB entry 1DLC which represents a good model for the target protein, it is not ranked in first place. This match cannot be identified by sequence comparison techniques (including PSIBLAST) and so is classed as a much harder problem than the previous example. In cases where no common ancestry is evident (i.e. superfold similarities) threading techniques, despite being somewhat unreliable, are the only means currently available for recognizing these similarities. (b) By shuffling the targets sequence randomly, a better estimate of the random background score distribution can be obtained. In this case, the best structural match for the target is clearly identified.

After running THREADER on the example sequence and ranking the results by column 13, which is the Z-score for the combined pairwise solvation energy, the results shown in Fig. 12.8 are produced. This is a typical result of an initial threading run for a difficult target sequence. In this case, one very close structural match for the target sequence (1dlc02) is included in the top 10 matches, but not in first place. In fact, the top five folds are all somewhat similar to the correct fold in that they are all β-sandwiches but are not sufficiently close to the native structure to allow a reasonable model to be built.

To better discriminate between good and bad models, THREADER allows the threading scores to be re-normalized against a random background. This is done by shuffling the targets sequence randomly and rethreading the shuffled sequence onto each of the candidate structures. The results of such a shuffling test is shown in Figure 12.7b. Figure 12.8 again shows the structures of both the target protein (Johnson *et al.*, 1996) and the best matching protein fold (Li *et al.*, 1991). Again, the proteins share a common fold, but with no obvious sequence similarity and no obvious functional similarity in this case, there is no evidence for common ancestry. Indeed, the fold in this case is one of the 10 known superfolds (Orengo *et al.*, 1994) and so no functional similarity can be inferred from the given structural similarity.

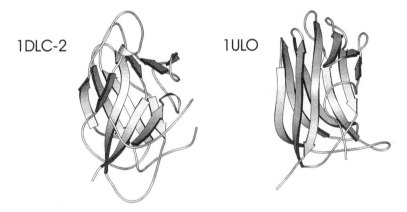

1DLC-2                1ULO

**Figure 12.8**    Molscript diagrams showing the structural similarity between the target protein (cellulose-binding domain from *Cellulomonas fimi*) and the matched database entry (1DLC).

## 12.12.2 Problems with threading

The single biggest difficulty that remains unsolved in fold recognition is probably how to deal with large multidomain proteins. It must be very clearly understood that most threading programs are aimed at recognizing single globular protein domains, and perform very poorly when tried on proteins which are far from this ideal. Threading a very large sequence, say 800 residues, is very unlikely to produce useful results unless the domain boundaries are known. Threading cannot be reliably used for identifying domain boundaries *at initio*. If the domain boundaries in the target sequence are already known, then the sequence should be divided into domains before threading, with each domain being threaded separately. Where predictions are attempted on very long multi-domain sequences then the results should be treated with suspicion, unless it is clear that the matched protein has a similar domain structure to the target. For example, the periplasmic small-molecule binding proteins (e.g. arabinose-binding protein) are two-domain structures (two doubly wound parallel α/β domains), but they all match each other fairly well as they all have identical domain organization. In contrast, however, pyruvate kinase has a number of quite distinct structural domains, and this organization is quite probably unique to pyruvate kinase. If a target is matched to pyruvate kinase without allowing for these distinct domains, then incorrect results will be obtained.

## 12.13   CONCLUSIONS

In this chapter a few key protein structure databases have been briefly reviewed and one popular application of these databases, namely the modelling of protein structure, has been described. It is important to realize that the new methods for protein structure prediction, for example threading, only became possible because of the existence of well-maintained databases of primary structural data. Most prediction methods are based on some kind of statistical analysis of structural data and this is really only possible when the data is stored in an organized form. The structural biology community were among the very first scientists to realize the importance of storing biological data in computer-readable form. The now somewhat 'infamous' Brookhaven PDB file format has been in existence since 1977, but although the format is often criticized today and a replacement format is already being promoted, many thousands of published papers in structural biology would not exist today if the crystallographers had not had the foresight to begin archiving their data so early on. In 1977, when the PDB was first officially announced it is well worth noting that this preceded the PC revolution by several years, and the recent 'Internet Age' by at least a decade.

As a concluding remark it is worth noting that as the several gigabytes of data in PDB can now be downloaded from the Internet and stored on the internal disk of a laptop computer in a matter of minutes it is easy to take it for granted. However, it must be noted that this data is the result of at least 1000 man (and woman) years of effort on the part of the experimental workers who collected the data and those who maintain the database itself (a sobering thought to bear in mind when you next find yourself spinning a protein structure around on a Web page somewhere).

## REFERENCES

Alexandrov N.N, Nussinov R. & Zimmer, R.M. (1995) In *Pacific Symposium on Biocomputing '96*, Hunter L. & Klein T.E. (eds). World Scientific Publishing Co., Singapore, pp. 53–72.

Altschul S.F., Gish W., Miller W., Myers E.W. & Lipman D.J. (1990) *J. Mol. Biol.* 215, 403–410.

Altschul S.F. Madden T.L., Schaffer A.A., Zhang J.H., Zhang Z., Miller,W. & Lipman D.J. (1997) *Nucl. Acids Res.* 25, 3389–3402.

Berman H.M., Olson W.K., Beveridge D.L., Westbrook J., Gelbin A., Demeny T., Hsieh S.H., Srinivasan A.R. & Schneider B. (1992) *Biophys. J.* 63, 751–759.

Bernstein F.C., Koetzle T.F., Williams G.J.B., Meyer E.F., Brice M.D., Rodgers J.R., Kennard O., Shimanouchi T. & Tasumi, M. (1977) *J. Mol. Biol.* 112, 535–542.

Bleasby A.J., Akrigg D. & Attwood T.K. (1994) *Nucl. Acids Res.* 22, 3574–3577.

Bowie J.U., Lüthy R. & Eisenberg D (1991) *Science* 253, 164–170.

Brown N.P., Orengo C.A. & Taylor W.R. (1996) *Comput. Chem.* 20, 359–380.

Bryant S.H. & Lawrence C.E. (1993) *Prot. Struct. Funct. Genet.* 16, 92–112.

Diamond R. (1992) *Prot. Sci.* 1, 1279–1287.

Fischer D. & Eisenberg D. (1996) *Prot. Sci.* 5, 947–955.

Flöckner H., Braxenthaler M., Lackner P., Jaritz M., Ortner M. & Sippl M.J. (1995) *Proteins* 23, 376–386.

Flores T.P., Orengo C.A., Moss D.S. & Thornton J.M. (1993) *Prot. Sci.* 2, 1811–1826.

Godzik A. & Skolnick J. (1992) *Proc. Natl. Acad. Sci. USA* 89, 12098–12102.

Grindley H.M., Artymiuk P.J., Rice D.W. & Willett P. (1993) *J. Mol. Biol.* 229, 707–721.

Hendlich M., Lackner P., Weitckus S., Floeckner H., Froschauer R., Gottsbacher K., Casari G. & Sippl M.J. (1990) *J. Mol. Biol.* 216, 167–180.

Hogue C.W.V., Ohkawa H. & Bryant S.H. (1996) *TIBS* 21, 226–229.

Holm L. & Sander C. (1993) *J. Mol. Biol.* 233, 123–138.

Holm L. & Sander C. (1997) *Nucl. Acids Res.* 25, 231–234.

Johnson P.E., Joshi M.D., Tomme P., Kilburn D.G. & McIntosh L.P. (1996) *Biochemistry* 35, 14381–14394.

Jones D.T., Taylor W.R. & Thornton J.M. (1992) *Nature* 358, 86–89.

Jones D.T., Miller R.T. & Thornton J.M. (1995) 23, 387–397.

Kraulis P.J. (1991) *J. Appl. Cryst.* 24, 946–950.

Laskowski R.A., Hutchinson E.G., Michie A.D., Wallace A.C., Jones M.L. & Thornton J.M. (1997) *TIBS* 22, 488–490.

Lemer C.M.R., Rooman M.J. & Wodak S.J. (1995) *Proteins* 23, 337–355.

Levitt M. (1997) *Proteins* S1, 92–104 (and other articles in same volume).

Li J.D., Carroll J. & Ellar D.J. (1991) *Nature* 353, 815–821.

Maiorov V.N. & Crippen G.M. (1992) *J. Mol. Biol.* **227**, 876–888.
Murzin A.G., Brenner S.E., Hubbard T. & Chothia C. (1995) *J. Mol. Biol.* **247**, 536–540.
Nussinov R. & Wolfson H.J. (1991) *Proc. Natl. Acad. Sci. USA* **88**, 10495–10499.
Orengo C.A., Flores T.P., Jones D.T., Taylor W.R. & Thornton J.M. (1993) *Curr. Biol.* **3**, 326.
Orengo C.A., Jones D.T. & Thornton J.M. (1994) *Nature* **372**, 631–634.
Orengo C.A., Michie A.D., Jones S., Jones D.T., Swindells M.B. & Thornton J.M. (1997) *Structure* **5**, 1093–1108.
Ouzounis C., Sander C., Scharf M. & Schneider R. (1993) *J. Mol. Biol.* **232**, 805–825.
Overington J., Donnelly D., Johnson M.S., Sali A. & Blundell T.L. (1992) *Prot. Sci.* **1**, 216–226.
Pascarella S. & Argos P. (1992) *Prot. Eng.* **5**, 121–137.
Russell R.B. & Barton G.J. (1994) *J. Mol. Biol.* **244**, 332–350.
Russell R.B., Copley R.R. & Barton G.J. (1996) *J. Mol. Biol.* **259**, 349–365.
Rypniewski W.R., Dambmann C., Vonderosten C., Dauter M. & Wilson K.S. (1995) *Acta Cryst. D.* **51**, 73–84.
Sali A. & Blundell T.L. (1990) *J. Mol. Biol.* **212**, 403–428.
Sowdhamini R., Rufino S.D. & Blundell T.L. (1996) *Fold. Des.* **1**, 209–220.
Taylor W.R. & Orengo C.A. (1989) *J. Mol. Biol.* **208**, 1–22.
Vath G.M., Earhart C.A., Rago J.V., Kim M.H., Bohach G.A., Schlievert P.M. & Ohlendorf D.H. (1997) *Biochemistry* **36**, 1559–1566.
Watson D.G. (1996) *J. Res. Natl. Inst. Stand. Technol.* **101**, 227–229.

# 13 PKR – the Protein Kinase Resource

*Michael Gribskov, Phil Bourne & Christopher M. Smith*

San Diego Supercomputer Center, San Diego, CA, USA

## 13.1 INTRODUCTION

Molecular biology is a leading discipline in the use of electronic data resources. Initially, this was due to the demand for efficient access to molecular sequence and structural data, but the use of databases and electronic archives has grown to include all areas of molecular biology. The driving force behind the rapid growth in electronic resources is the massive amounts of detailed information available on metabolism, regulation, intracellular and intercellular interactions, and all of the other minutiae of cellular and molecular biology. What in 1970 Watson could describe in his single volume *Molecular Biology of the Gene* now requires an entire library. To deal with this enormous growth in biological knowledge, scientists are increasingly turning, almost out of desperation, to computerized database management systems to index and cross-reference this information so that it is usable. While this effort is in its early stages, one can foresee that in the near future most of the biological literature will be available in electronic form. The electronic literature will be able to directly draw on increasing amounts of reference material defining the structures, sequences, functions, and genetics of molecular entities.

Many of the first-generation of molecular biological data resources were targeted at broad coverage of a certain type of data: GenBank (Benson *et al.*, 1996) and EMBL (Stoesser *et al.*, 1997) cover all nucleic acid sequences, SWISS-PROT (Bairoch & Apweiler, 1997) and PIR (George *et al.*, 1996) cover all protein sequences, and PDB (Abola *et al.*, 1987) covers all three-dimensional structures. These broadly targeted data resources have gradually become highly efficient at collecting and archiving their respective data, and are continually improving their abilities to retrieve specific information in a user-friendly manner. In general, these efforts have been able to become successful only by severely limiting the amount of curation they perform on the

*Genetics Databases*
ISBN 0-12-101625-0

data. Curation requires highly trained experts to examine the data and, where necessary, correct it or add pointers to related information. Curation is thus both time-consuming and costly, and in the face of a continual exponential growth in the data, has had to be automated or curtailed. These first-generation databases may therefore be characterized as broad but shallow. They cover all information of a certain type, but provide only a limited amount of additional information or references to related information.

Over the last decade, there have been many efforts to create a second generation of databases, drawing on the lessons and, more importantly, the information of the first. Many of the efforts attempted to integrate the information of the first-generation databases into monolithic databases. This proved to be enormously difficult, akin to trying to introductory integrate texts on physics, biochemistry and chemical engineering into a single volume. More successful efforts attempted to define narrower topics, such as human genetics, and to add detailed annotation to the information present in the first generation databases. Many of these efforts are successful, for example, OMIM (McKusick, 1994), KEGG (Kanehisa, 1997) and many others. A lesson learned from this work was that it is difficult to completely integrate the primary databases because of differences in how the databases define their data, and because of the ongoing development of new scientific concepts. This has led to the concept of 'federation' – databases that can communicate with each other, but still maintain their own view of the data and how it is defined.

A third generation of database is currently being constructed. These data resources are typically even more narrowly focused, often targeting only a single family of proteins, or a specific set of genes. Scientists working in the area targeted by the database, thus providing in-house expertise and annotation usually construct them. And they attempt, as much as possible, to take advantage of the information and capabilities of the first generation databases without duplicating their data. This kind of database can be contrasted with the broad and shallow databases as being narrow but providing a greater depth and richness of annotation and linkage to primary resources. The Protein Kinase Resource (Smith *et al.*, 1997) at SDSC is an attempt to create such a deep but narrow database that, hopefully, can be used as a prototype for many similar databases.

## 13.2   DATA DEFINITION

PKR takes a federated database approach. That is, the goal is to only maintain local copies of the data needed to provide local services. An example is the protein sequences stored locally in order to search and compare the sequences. However, PKR does not copy the entire entry from the source database because information such as deposition dates, literature references, taxonomy and associated annotation may be conveniently retrieved directly from the source database using a web link. The difficulty lies in keeping the local data in step with changes in the public resources. For example, the names of sequences in SWISS-PROT frequently change as new sequences are entered into the database. PKR must recognize the change and update its data in a coordinated way. The key to this approach lies in detailed data modelling. PKR uses data definitions written in the STAR data definition language (Hall, 1991; Hall & Spadaccini, 1995) to model the data items included in the resource. There are currently five dictionaries that support the main types of data represented in PKR. Three-dimensional structural data is represented using the mmCIF dictionary (Bourne *et al.*, 1997), a public standard for three-dimensional coordinate data. Sequences and sequence features are represented using new dictionaries developed for PKR. The sequence dictionary defines data items that pertain to individual sequences such as the amino acid sequence itself, the enzyme name, and locations of known motifs and features within the sequence. The features dictionary defines sequence and structural motifs that are found in many different structures. This dictionary allows one to define a motif, such as a kinase catalytic domain, and to specify in which sequences it is found and how the various instances of the motif in different sequences map onto each other. Figure 13.1 provides a simple example of the definition of a data item, in this case, the name of a sequence in the source database in the sequence dictionary. A fourth dictionary represents data for enzymology such as reaction rate constants, equilibrium constants, etc. The fifth dictionary defines common items used by many of the other dictionaries such as a common syntax for literature citations.

```
save__sequence.name.
_item_description.description
;
     Trivial name of this entry in the source database. Trivial names
     are subject to change without notice by the database mantainer.
;
_item.name                '_sequence.name'
_item.category_id         _item.category_id
_item.mandatory_code      no
_item_type.code           line
_item_examples.case       'ABL_DROME'
```

**Figure 13.1**   Example of data dictionary: _item_description.description gives a text description of the data item; _item.name defines the name of the data item, i.e., 'sequence.name'. The item.mandatory_code specifies whether this data item is required and _item_type.code specifies that this data is of type 'line', which is defined elsewhere as up to 80 characters of alphanumeric text. Finally, _item_examples.case gives an example of a valid sequence name.

## 13.3    PKR FUNCTIONS AND DATA

Serine-threonine and tyrosine protein kinases are an important class of molecules that play important roles in intracellular and intercellular signalling. A great deal of information is available for these enzymes in electronic form, including, three-dimensional structures, protein and nucleic acid sequences, gene locations, genetic diseases, and a growing body of detailed information on specific molecules.

### 13.3.1    Sequence information

There are now approximately 4000 known serine/threonine and tyrosine kinases or kinase-like sequences (see http://www.sdsc.edu/Kinases/pkr/pk_catalytic/pk_cat_list.html). This list is updated monthly by scanning the protein sequence databases with a panel of catalytic regions representing all families of serine/threonine and tyrosine protein kinases. An initial alignment of 398 kinase catalytic regions provided by Hanks & Quinn (1991) has been expanded to include nearly all of these actual and putative kinases. The kinases are organized into five major groups: the cAMP/cGMP dependent kinases, $Ca^{2+}$/calmodulin regulated kinases, CMGC kinases, tyrosine protein kinases, and other undefined types. These five major groups are further divided into 55 subgroups by a combination of sequence and functional criteria. Multiple sequence alignments for each of these groups and subgroups can be displayed in a variety of formats, including colouring schemes that highlight the conservation and chemical similarity of aligned residues. Interactive tools that permit the calculation of summary statistics, e.g. residue frequencies at specific positions for user-selected groups of sequences, and that will analyse and assign novel sequences to the known classification are under development. Additional multiple sequence alignments describing kinase-associated domains such as SH2 and SH3 domains, and calcium binding domains are also under development. All sequences are cross-referenced with public databases such as SWISS-PROT (Bairoch & Apweiler, 1997) and PROSITE (Bairoch *et al.*, 1997), PIR (George *et al.*, 1996), GenBank (Benson *et al.*, 1996), EMBL (Stoesser *et al.*, 1997), and PDB (Abola *et al.*, 1987). Ultimately, the sequence information will be cross-referenced with the three-dimensional coordinate information so that sequences of interest can be immediately associated with the relevant three-dimensional structures.

## 13.3.2 Structural information

Three-dimensional structural information is available for 36 distinct structures. Detailed information available for each structure includes a brief synopsis of the structure, bibliographic references, crystallographic information and, for major members of the family, a structure walk-through that details the structure and function of the molecule. The long-term goal of the walkthrough is to provide a level of annotation and explanation of structural features that is not possible within the page constraints of traditional journals. Beyond a discussion of each individual structure, PKR provides an opportunity to compare structures via collages showing contact matrices, isotropic temperature factor distributions, and $C_{\alpha}$ superpositions exhibiting the open and closed forms of the enzyme (Zheng *et al.*, 1993). Three-dimensional structures may be rendered by the QuickPDB Java applet (Shindyalov & Bourne, 1998), as VRML, using RasMol (Sayle & Milner, 1995), or using the Molecules R Us facility (http://molbio.info.nih.gov/doc/mrus/mol_r_us.html). PKR also provides functions for comparing kinase structures and structural elements and for viewing structure-based alignments of kinases (see http://www.sdsc.edu/Kinases/pkr/pk_structure.html#Analyses).

## 13.3.3 Genetic information

PKR contains over 400 links to OMIM (McKusick, 1994) entries detailing kinase-related genetic diseases. Much of this information is in textual form, so a major effort will be required to analyse the links and limit them only to those related to serine/threonine and tyrosine protein kinases. As additional genetic information becomes available, such as chromosomal locations of genes, it will be incorporated into PKR.

## 13.3.4 Ancillary information

The goal of PKR is to present information that is useful to laboratory scientists. In addition to the detailed molecular information described above, PKR therefore maintains a number of additional links and resources of use to the kinase community. These include a list of protein kinase–protein phosphorylation workshops and conferences, listings of protein kinase researchers, and pointers to reagents and DNA/protein synthesis/sequencing facilities. PKR accepts and distributes laboratory protocols used in the kinase community, including purification methods, enzymatic assays, and others. An active discussion group (kinases@sdsc.edu) in which kinase researchers contact each other with questions and information is hosted by PKR using a list server.

## ACKNOWLEDGEMENT

This project is supported by the National Science Foundation through Cooperative Agreement ACI 9619020 to the National Partnership for Advanced Computing Infrastructure and grant NSF-9630339 to P. Bourne, I.N. Shindyalov and M. Gribskov.

## REFERENCES

Abola E.E., Bernstein F.C., Bryant S.H., Koetzle T.F. & Weng, J. (1987) In *Crystallographic Databases – Information Content, Software Systems, Scientific Applications*, Allen F.H., Bergerhoff G. & Sievers R. (eds). Data Commission of the International Union of Crystallography, Bonn/Cambridge/Chester, pp. 107–132.

Bairoch A. & Apweiler R. (1997) *Nucl. Acids Res.* 25, 31–36.

Bairoch A., Bucher P. & Hofmann, K. (1997) *Nucl. Acids Res.* 25, 217–221.

Benson D.A., Boguski M.S., Lipman D.J. & Ostell J. (1996) *Nucl. Acids Res.* 24, 1–5.

Bourne P.E., Berman H.M., McMahon B., Watenpaugh K.D., Westbrook J. & Fitzgerald P.M.D. (1997) *Meth. Enzymol.* 277, 571–590.

George D.G., Barker W.C., Mewes H.-W., Pfeiffer F. & Tsugita A. (1996) *Nucl. Acids Res.* 24, 17–20.

Hall S.R. (1991) *J. Chem. Inf. Comput. Sci.* 31, 326–333.

Hall S. R. & Spadaccini N. (1994) *J. Chem. Inf. Comput. Sci.* 31, 505–508.

Hanks S. & Quinn A.M. (1991) *Meth. Enzymol.* 200, 38–62.

Kanehisa M. (1997) *Trends Genet.* 13, 375–376.

McKusick V.A. (1994) *Mendelian Inheritance in Man. Catalogs of Human Genes and Genetic Disorders.* Johns Hopkins University Press, Baltimore.

Sayle R.A. & Milner E.J. (1995) *TIBS* 20, 374–376.

Shindyalov I.N. & Bourne P.E. (1998) QuickPDB. http://www.sdsc.edu/pb/Software. html.

Smith C.S., Shindyalov I.N., Veretnik S., Gribskov M., Taylor S.S., Ten Eyck L.F. & Bourne P.E. (1997) *TIBS* 22, 444–446.

Stoesser G., Sterk P., Tuli M.A., Stoehr P.J. & Cameron G.N. (1997) *Nucl. Acids Res.* 25, 7–13.

Watson J.D. (1970) *Molecular Biology of the Gene.* W.A. Benjamin, Inc., Menlo Park.

Zheng J., Knighton D.R., Xuong N-H., Taylor S., Sowadski J.M. & Ten Eyck L.F. (1993) *Prot. Sci.* 2, 1559–1573.

# 14  Gene Expression Databases

## Richard Baldock & Duncan Davidson

MRC Human Genetics Unit, Western General Hospital, Edinburgh, UK

## 14.1  INTRODUCTION

The Human Genome Mapping Project and similar sequencing projects are providing a huge quantity of information about the structure of the genome. The
next, and perhaps more challenging, task is to understand the functions and
interactions of the genes through all levels from cellular expression to mor-
phogenesis and normal and abnormal phenotype. Studies of gene function at
the cell and tissue level typically rely on evidence from the effects of mutation
or modified expression and gene-expression assays. One step to understanding
how these genetic networks can control more macroscopic events is to detect
the location and abundance of gene-products as a measure of gene activity
within the tissues of interest. The comparison of the spatial organization of
gene activity with the activity of other genes, anatomy, histology, metabolic
activity, etc., may provide important clues to the genetic control mechanisms
and lead to further experiment as well as provide baseline data for models of
the underlying biology.

Useful information for this enterprise can come from a whole range of
experiments and organisms. As more and more expression patterns are being
reported it is becoming increasingly useful to scan the data for genes with overlapping or complementary expression in order to frame novel hypotheses
about genetic interactions that might subsequently be tested by molecular
genetic experiments.

In response to this requirement a number of gene-expression databases have
been set up (Table 14.1); most are in the early stages of their development and
the list will undoubtedly expand. Some of these databases have an underlying
database management system (DBMS) and can provide an interface to allow a
range of search facilities, others are a series of World Wide Web (WWW)
hypertext pages linked 'by hand' as new data is entered. In all cases, the means

*Genetics Databases*
ISBN 0-12-101625-0

**Table 14.1** WWW addresses of gene-expression and related databases. These addresses are also accessible from http://genex.hgu.mrc.ac.uk/Urls/. Each column indicates (Y, yes; N, no) whether the database holds that type of gene-expression data.

| Database name | WWW address (note these addresses should be preceded with http://) | Gene name | In situ | Homologue | Time | Space[1] | Quantification | Gene product | Submission | Internal links | External links | Comments |
|---|---|---|---|---|---|---|---|---|---|---|---|---|
| NEXTDB | watson.genes.nig.ac.jp:8080/db/ | Y | Y | N | Y | T,P | N | N | Y | N | Y | cDNA clones, cosmids |
| ACeDB | www.sanger.ac.uk/ | Y | Y | Y | Y | T,P | N | N | N | Y | Y | |
| | www.personal.leeds.ac.uk/~acedb/Hope/epa.htm | N | Y | N | Y | T,P | N | N | N | Y | N | cosmids |
| Flyview | flyview.uni-muenster.de/ | Y | Y | N | Y | T,P | N | N | Y | Y | Y | enhancer trap lines |
| The Flybrain Project | flybrain.uni-freiburg.de/ | Y | Y | Y | Y | T,P,M | N | N | N | N | Y | |
| Flybase | flybase.bio.indiana.edu/ | Y | N | N | Y | T | N | Y | Y | Y | Y | |
| The Zebrafish Database | zfish.uoregon.edu/ | Y | Y | Y | Y | T,P,A | N | Y | Y | Y | Y | GE not yet available |
| The Xenopus Molecular Marker Resource | vize222.zo.utexas.edu/ | Y | Y | Y | Y | T,P | N | Y | Y | Y | Y | |
| The Mouse Gene Expression Information Resource | genex.hgu.mrc.ac.uk/ www.informatics.jax.org/ | Y | Y | Y | Y | T,P,M, T,D,A | Y | Y | Y | Y | Y | not fully available |
| TBASE | tbase.jax.org/ | Y | N | N | N | T | Y | Y | Y | Y | Y | |
| dbEST | www.ncbi.nlm.nih.gov/dbEST/index.html | Y | N | N | N | N | Y | Y | Y | N | Y | |
| The Kidney Development Database | www.ana.ed.ac.uk/anatomy/kidbase/kidhome.html | Y | Y | Y | Y | T,P | N | N | Y | N | Y | links to refs |
| The Organogenesis Database | www.ana.ed.ac.uk/anatomy/kidbase/orghome.html | Y | Y | Y | Y | T | N | N | Y | N | Y | links to refs |
| The Tooth Gene Expression Database | honeybee.helsinki.fi/toothexp/toothexp.html | Y | Y | Y | Y | T,P,D | N | N | Y | Y | Y | |
| The Ion Channel Database | pain.med.umn.edu/csn/ | Y | N | Y | N | T | Y | Y | Y | Y | Y | |
| The Human Anatomy Database | ed.ac.uk/anatomy/database/humat/ | N | N | N | Y | T | N | N | N | N | N | anatomy nomenclature |

of access is restricted to the interfaces provided by the database curators. As a consequence of the relative newness of these databases and the fragmented and varied approach to their design, the functionality is fairly limited by comparison with sequence and gene-mapping databases. In attempting to use expression databases to address questions of gene function, the biomedical community has a challenging task ahead.

To begin to meet this challenge, the prospective user of gene-expression databases needs to know what types of information they contain, how to scan the data, and how to interpret the results and move forward to further investigations of gene function. One problem in writing about gene-expression databases is that, for the few that are currently available, operation is simple and self-explanatory: the more complex, planned databases are not yet operational. In this chapter, therefore, we aim to give a brief comparative overview of present databases and to identify issues that arise from the nature of the data and that must be addressed in the future. We begin by providing a brief synopsis of the types of gene expression assay that are represented followed by an outline of the range or scope of existing databases, including those under active development. We then discuss the data requirements and provide a brief review of currently available data. Finally, we discuss issues that need to be addressed and likely future directions in the development of these databases as bioinformatics resources.

The databases discussed in this chapter are evolving rapidly. The best way to use this guide is to get an overview of the various resources and then to sit at a workstation and access the databases directly using the addresses in Table 14.1. Alternatively, the table of database links given on the Mouse Atlas and Gene Expression Database pages (http://genex.hgu.mrc.ac.uk/) may be used.

## 14.2   GENE-EXPRESSION ASSAYS

The expression of a gene can be established or inferred in a number of ways. The initial concept of a gene (Mendelian gene) and evidence for its expression arises from studies of phenotype and inheritance. With the ability to sequence the genome, new evidence for genes (genomic genes) and, by implication, gene expression, comes from the existence of possible gene sequences (e.g. open reading frames, ORF, and conserved sequences). This is not the subject of the present chapter. Here, we are concerned with assays that provide direct evidence of the activity of a gene by detecting one of the products in the process of gene expression. Three key aspects of the assays can be identified and used to classify the type of the experiment.

■ Gene product detection:
  ● hybridization – unprocessed or processed RNA
  ● immuno-chemistry – protein
  ● tagging – expressed-sequences (RNA) and cDNA
  ● RNA-ase protection – RNA
  ● RTPCR-RNA
■ Spatial/tissue resolution:
  ● high – *in situ*-sections, whole-mount, cells
  ● low – homogenized dissected tissue, organ culture, cell lines, sorted cells
■ Quantification (molecular species abundance):
  ● quantitative – Northern blot – RNA
  ● quantitative – Western blot – protein
  ● semi-quantitative – dotBlots (nonsegregated)
  ● nonquantitative – *in situ* hybridization
  ● semi-quantitative – array (chip) hybridization – RNA

Techniques that require homogenized material cannot give good spatial resolution unless the tissue has been dissected very carefully or the cells sorted in some way. Even if this has been done these assays cannot provide information of the gradients and shape of the gene-expression patterns in the original material: this is only available from *in situ* methods. The quantification that is available using gels and blots is the number of target molecules relative to a standard preparation. This may not be easily converted back to an expression density in the original tissue. In this sense *in situ* methods are nonquantitative and cannot accurately provide measures of relative abundance between adjacent sections for the same target molecules – even less the relative abundance of different molecules that could be probed. *In situ* methods do, however, provide a good measure of relative abundance within a given preparation and therefore useful information on expression gradients, 'hot-spots' and other distribution-pattern features. An important point to keep in mind is that all of these assays are measuring a snapshot of a dynamic system and the strength or numbers of a molecular species detected is dependent not only on the rate at which the gene is being transcribed but also on the rate at which the product is utilized and cleared.

Finally, new techniques are always being developed, for example probes that can be localized using magnetic resonance imaging (MRI) (Ahrens *et al.*, 1997) will ultimately allow real-time measurement of the spatial distributions and strength *in vivo*.

## 14.3  DATABASE SCOPE

Gene-expression databases have been developed to reflect the interests of the associated groups, sometimes as an extension to existing systems holding, e.g.

sequence and mapping data (ACeDB). Alternatively, the interest is in specific organs across many species (e.g. tooth and kidney). Others have been designed to record gene expression in cells *in vitro*, embryo and the adult organism (e.g. zebrafish, mouse). All hold data with respect to wild-type and inbred strains and can, in principle, contain information with respect to mutants and disease states though direct spatial mapping of the data may be more difficult. TBASE concentrates on transgenic and mutant organisms (currently *Drosophila*, mouse, rat and pig).

Not all the gene-expression databases hold detailed descriptions or images of the gene-expression patterns but they can provide contact addresses or cite a publication. This is a useful but limited view that cannot provide a mechanism for many types of query. More complete systems are being developed which will record all the data necessary for queries that involve complex spatial and temporal constraints as well as links to other databases with sequence, mapping and bibliographic data.

For some databases, gene-expression data is only a small part of the information held. A good example of this type of database is ACeDB which contains genome sequence, genetic and physical mapping data. Another example is the Zebrafish Database that will contain a wide range of phenotypic and genetic data as well as contact addresses, etc.

At least one of the gene-expression databases (MGEIR) will have pattern data mapped onto a standard spatial reference, or *gold-standard*, embryo for each of the developmental stages that will enable specifically spatial queries to be made e.g. 'Which expression patterns coincide?'. 'What is expressed within a certain spatial range?', and 'What other genes show the same or complementary distribution of expression (in terms of shape)?'.

The full potential of these gene-expression resources will only be realized when the user has the option to pose complex queries, not just with spatial and temporal relationships, but also with links to sequence, mapping and functional data. These interfaces are not yet available, although progress in the area of interoperability is being made for other bioinformatics resources such as TAMBIS (Transparent Access to Multiple Biological Information Sources, http://www.cs.man.ac.uk/mig/tambis/). To date, however, the query facilities for existing gene-expression databases are primitive and need to be extended to include the space–time constraints in addition to key terms about the experiment (e.g. gene, probe, strain, etc.). Examples of these constraints are: 'which genes are expressed in tissue A and not in tissue B?', 'which genes are expressed in the regions where genes X and Y are co-expressed?' and 'which genes are expressed in tissue A *after* the expression of gene X?'. More complex spatial queries could include shapes of the gene-expression distributions: 'which genes exhibit the same or complementary pattern of expression to that of gene X?'. This shape could also include the expression intensity profile: 'which genes show the same expression gradients as gene X in tissue A?'.

These queries will become possible as the databases develop. Already some have links to other databases and references to more complete information, as well as pointers to genetic and biochemical evidence relevant to gene function.

## 14.4    GENE-EXPRESSION DATA

In future, the gene-expression databases will be part of a range of bioinformatics resources to help understand gene function. Interoperability between these resources has not yet been implemented but a prerequisite for comparison is that databases hold homologous information and ideally an agreed minimal set for such purposes. This is not yet the case (see Table 14.1) but as standard mechanisms for accessing complex data, in particular the Common Object Request Broker Architecture (CORBA; Object Management Group 1995), are adopted then the incentive to include more complete data-sets will increase.

In this section, aspects of this minimal set of data are discussed and for each database summarized in Table 14.1. A key issue in database design is that to prevent inconsistencies in the data, duplication should be avoided. Rather, links to the appropriate primary source should be included. Another general point is that where possible original data should be entered rather than interpretations so that, if necessary, the data can be reassessed by the user.

### 14.4.1    Gene name

Most databases use standard gene names that are recognized by the appropriate nomenclature committees. The use of standard text for gene names and symbols is important for searching and interpreting the data and provides a key link for relating information in different databases (e.g. sequence, mapping and function).

### 14.4.2    Method/assay

Sufficiently detailed information on the assay method used to obtain the information is required to fully understand the data in terms of specificity, sensitivity and resolution in space and time. Few of the databases cited provide all this information. Sufficient detail of the probe or antibody is required to assess the specificity, i.e. what has been detected – for example, does the data distinguish between different splice forms or gene products?

Sensitivity relates to what levels of gene expression can the assay distinguish from a background response. Critical aspects of the assay, e.g. hybridization and wash temperatures, salt concentrations, etc., may also be necessary for interpreting the specificity of the data. Databases should be encouraged to include this information as an annotation if the author of the data decides that this information is important for that interpretation.

Resolution is the accuracy with which expression patterns can be delineated either in space or time. The blot assays are more sensitive and *in situ* methods have a higher resolution. Information on whether *in situ* assays were carried out in sectioned material or in whole-mount preparations is also important since this affects the expected resolution of the data. It is more difficult in whole-mount preparations than in sectioned material to determine which tissues display gene expression.

It is clearly also relevant to record which tissues were examined but in which no expression was detected, as distinct from tissues or stages of development that were not examined. Most databases do not include this information.

## 14.4.3   Temporal information

Substantial amounts of gene-expression data are recorded with respect to development and links between databases in terms of developmental stage are potentially very important. For most organisms standard staging terms related to embryo development have been defined (e.g. Theiler stages in mouse, Carnegie stages in human) but most data is entered using day post conception (d.p.c.) which is less accurate because individual embryos of the same age may vary considerably in developmental stage. Ideally, these should be entered using the staging systems that are often more accurate and meaningful for temporal comparisons and can provide some degree of inter-species comparison.

In some cases, multiple staging systems have been defined for finer temporal resolution (e.g. Theiler stages and somite count in mouse), so it should be possible to record these additional details as part of the gene-expression entry. Ultimately, more-accurate staging mechanisms may emerge. For example, molecular markers for specific cellular events (e.g. muscle cell differentiation) could be important. Ideally, it should be possible to record expression as 'before, concurrent with, or after', the onset of expression of a marker gene as assayed in the same experiment.

## 14.4.4   Spatial information

The spatial patterns of gene expression can be recorded in text or graphical form. Text is used to describe the location of the expression via a controlled vocabulary such as anatomical nomenclatures (e.g. mouse, human, zebrafish,

Flyview, Flybase), free text (e.g. the *Xenopus* Molecular Marker Resource and TBASE) or by identifying individual cells. In *Caenorhabditis elegans* there is a unique means of identifying cells on the basis of lineage. The MGEIR will include an anatomical description of the mouse embryo that is organized in an open-ended database to which finer detail can be added as required. The mouse anatomical nomenclature is being used, where appropriate to help formulate descriptions of zebra fish and human embryos. It is to be hoped that the use of standard nomenclature systems will not only simplify the use of each of these databases, but will enable cross-database links and queries.

Graphical methods include diagrams or line drawings, 'snapshot' pictures of the data and image data mapped onto a standard reference frame. In the Tooth Database, gene expression is assigned to tissues that are represented in diagrams of tooth primordia at successive stages of development. This reduces confusion of the use of different names but data can not be assigned to subregions of a tissue. In MGEIR, some diagrammatic annotation is employed to qualify the approximate section position and orientation.

In addition to diagrammatic annotation, many databases hold images that illustrate the details of complex gene-expression patterns. A major distinction here is between those databases that document the expression of different genes in a series of independent or snapshot images and those in which different gene-expression patterns are spatially mapped onto a set of standard reference images (two- or three-dimensional). Spatial mapping allows immediate comparison between different patterns and, in principle, allows purely spatially constrained search. The main advantage of spatially mapped data is, however, that many gene-expression patterns do not map $1:1$ with an anatomical description so that gene-expression data often cannot be translated simply into a list of anatomical terms. This is particularly true of genes involved in mechanisms that operate in a spatial, rather than tissue-specific, context during the early stages of development, when the anatomy of the organism is being established. The spatial mapping approach has considerable potential since it can link gene-expression data to other spatially distributed information that is relevant to gene function, for example, cell lineage, cell proliferation, cell death, etc. A disadvantage is that the data is transformed from its original form. This is potentially more laborious, although possibly more objective than simply cataloguing original images; moreover, the mapping must be done carefully to ensure that significant features are not lost or spurious features introduced. The *Drosophila* database being developed at the University of California (Hartenstein *et al.*, 1995), the Zebrafish Database and the MGEIR plan to employ spatial-mapping.

In direct analogy to the temporal markers, as more gene-expression patterns are entered into databases it is likely that some will become established as spatial reference patterns. These will be patterns that are well defined, invariant, and dynamically stable, and for which probes or antibodies are readily available. These gene-expression domains will define a 'molecular anatomy' against which the expression of other genes can be described. Such a molecular anatomy may reflect a developmentally more profound subdivision of the embryo than the morphological subdivision that is presently recognized by classical anatomy.

Finally, when whole-mount or section data is entered into the data base it is important to enter the gene expression pattern for the entire region examined. This implies that the regions examined and considered not to express the gene are entered as 'not expressed' as well as regions of positive expression. This is important so that the user can distinguish between an observation of no expression and not observed.

## 14.4.5 Quantification

Methods applied to homogenized material, for example reverse transcription polymerase chain reaction (RT-PCR) and 'chip'-based methods, can potentially provide quantitative information on the numbers or relative numbers of the molecular species that are detected. At present, none of the databases record quantitative differences in gene expression, but since these differences in the concentration of gene product are important, quantitative data is likely to be recorded in the future.

Techniques for assaying gene expression *in situ* are not quantitative, except in the sense that semiquantitative comparisons can be made within the same whole-mount preparation or histological section. The best that can be done is a simple weak/moderate/strong annotation of the *in situ* signal or the inclusion of representative images to illustrate, for example, gradients of expression. Some planned databases (for example, the MGEIR) will record textual information on quantitative aspects of gene expression and will allow the annotation of gene expression 'hot-spot'. As the databases become more detailed the gene-expression information recorded will include the relative expression intensities within the *in situ* sample and therefore expression gradients can be determined.

## 14.4.6 Gene products

In many cases, detailed knowledge of the nature of the gene product detected in the experiment may only be available from other data. For example, evidence of gene expression in tissue extracts assayed by electrophoresis may provide additional information on product size or post-translational processing. In this case, it is important that the *in situ* data is associated with the additional information generated by these other assays (see 14.4.8). Few gene-expression databases currently store such data.

## 14.4.7    User annotation of existing data

Annotation of existing entries by other users of the database could be a very valuable source of additional information, especially for users not familiar with that gene or organism. It could also add to the interpretation of the data and provide a means for additional links. None of the current gene-expression databases provide for annotations to the data by other users.

## 14.4.8    Linked entries

Entries in existing gene expression databases relate to individual experiments and individual genes. In many cases, experiments are performed to establish coincidence (or non-coincidence) of expression of multiple genes or gene products or have additional analysis of the gene products. Knowing that two assays were performed at the same time on the same material can provide much more confidence that the relationship between the assays is real rather than chance overlap. Such links may also be important to describe gene expression by comparison with data on controls. Gene-expression databases do not currently provide the possibility of linking different entries. None of the currently available databases deals explicitly with control data. At least some of the databases that are now being developed (for example, the MGEIR) plan to address these issues.

## 14.4.9    Links to other databases

Relationships between data in different databases can be established by query provided that there is commonality of meaning of terms or key words. This type of direct interoperability is not available for existing gene expression databases and a simpler mechanism for marking significant relationships is to include direct links between databases. These are entered 'by hand' and are static but can provide useful clues when browsing the database entries. There is considerable potential for links between different gene-expression databases relating to the same organism, for example in the mouse between the Tooth Database, the Kidney Development Database and the MGEIR.

Links between gene-expression databases dealing with different species will also be important. This is particularly the case between mouse and human (Davidson & Baldock, 1999), but links between mouse, zebrafish and *Drosophila* databases will also be important. Genetic studies in these organisms are providing an important starting point for identifying mammalian genes.

As we have discussed briefly in Section 14.3, links from gene-expression databases to databases dealing with sequence, genetic linkage, phenotypic and mutant data will also be important in using evidence concerning gene expression in studies of gene function. An additional way to link related information from different databases is to abstract data and combine them in a single resource aimed at a particular audience. An excellent example of this approach is the Ion Channel Database (Table 14.1).

All of these links depend on the compatibility of data (particularly on the use of standard nomenclature for genes and anatomical components, etc.), the feasibility of relating data models, the practicality of inter-database links at the software level and the motivation across the community to fund and implement the necessary work. At present, none of the gene-expression databases are linked in this way, although the need for integration is appreciated.

## 14.5 ACCESS AND SUBMISSION

### 14.5.1 Access

All the databases described here are accessible via the WWW. The minimum requirement to access them is a computer with an adequate connection to the Internet: this may be a Macintosh, IBM-type PC running Windows, or a Unix workstation. A WWW browser such as Netscape Navigator or Microsoft Internet Explorer is also required. In some cases, up-to-date versions of the browsers are required to handle the frames and tables or with more recent versions of Java. This may exclude the use of Macintosh and some Unix machines. In each case, the database instructions provide details of the machine and browser functionality required for access.

Several databases currently also provide improved data interfaces using Java either as *applets* or *applications* (standalone Java programs). Applets are small programs that run inside the web browser and their use requires a browser that can interpret Java. Java applications require a Java interpreter to run the program, these are free and fully up-to-date for MicroSoft Windows and Solaris (Sun Microsystems) workstations but may not be available for Macintosh and other Unix systems.

### 14.5.2 Submission

Not all of the databases described here accept submissions (see Table 14.1). Those that do provide more or less detailed instructions on their respective WWW pages; here, we provide a brief overview of how different databases deal with data submission. An important distinction is between those that accept on-line data submission by individual users and those in which data is entered centrally by the database team or editors. These latter type's data may either be from literature searches (as for example, in the MGEIR) or from experimental results sent by individual laboratories by *ad hoc* mailing, as, for example in the Flyview database. Some databases plan to operate both systems (e.g. the Zebrafish Database and the MGEIR).

Gene-expression databases are recent developments in the field of bioinformatics and therefore the data are sparse and incomplete. This is mainly because relatively few users can devote time to data submission; thus only a few per cent of entries are submitted by users, the rest come from literature searches by the database teams. Exceptions to this are the smaller databases (e.g. Kidney Development Database) for which an active core of users submit data and the database has reasonably comprehensive coverage.

A key issue in any database is quality control, especially if any of the data is interpreted and the original data is not accessible. In the case that the data has been abstracted from peer-reviewed journals there is a clear standard. Other cases have a more or less *ad hoc* approach. For example, MGEIR plans to request that *in situ* data is accompanied by at least one image of original data to enable quality to be assessed by the user.

The *raison d'être* for these database systems is to hold data that cannot be made available and be queried in some other way. Quantity of data is not a limitation and a good gene-expression database can hold much more extensive and detailed information than can be published in a conventional paper. Each entry may comprise only a small amount of data. For example, a single entry of data from an extensive *in situ* hybridization screen may include information from only a single section assayed using an uncharacterized probe. This information could be of direct interest to other groups and therefore acceptable, provided that the data quality is adequate and that the probe is made available to those for whom the effort of further characterization would be worth while.

To interpret the results of a query it is necessary to understand how the data was entered into the database. A good example of this is how to interpret the result of a spatial query such as 'what genes are expressed here (a region selected by delineating part of the image using the computing pointing device e.g. a mouse or graphics tablet)?'. If the answer is none, then the user must be able to distinguish between the case when there are no entries near the probe region and the case when there are entries that show that the region has been studied and no expression was observed. In this case the user must realize that it is sensible to ask 'what genes have not been observed here?'. It is also important to realize that most databases are far from comprehensive and that much of the data is incomplete and some may be inaccurate or entered only at a low resolution. In this light, gene-expression databases should be seen, not as repositories of truth, but as tools by which to frame hypotheses that should then be tested by experiment.

## 14.6  SPECIFIC DATABASE SYNOPSES

In this section we discuss specific databases that are currently available or under development and indicate how they can be accessed. This information is up-to-date at September 1998.

### 14.6.1  Nematode – *Caenorhabditis elegans*

There are a number of databases for the nematode that are based on the database system ACeDB, and one is available as a set of WWW pages that may be searched.

#### *14.6.1.1  NEXTDB*

The Nematode Expression Pattern Database (NEXTDB) is the beginning of an expression pattern map of the 100 Mb genome of *C. elegans*. The genome is being mapped through clustering of random cDNA clones by tag sequencing and the sequences expressed are being analysed by *in situ* hybridization on whole-mount embryos, larvae and adults. The data available are: maps of the relationships between cosmids, cDNA clones and the genes (predicted), images of the *in situ* hybridization results, sequences of the cDNA clones and the results of BLASTX search for homologies. This is a database under construction, with future information to include image annotation particularly with identification of the cells showing positive expression at each stage and more detail of the cDNA clustering. The database is accessible via the WWW with any browser.

#### *14.6.1.2  ACeDB systems*

A *C. elegans* Database (ACeDB – Martinelli *et al.*, 1997) is somewhat different from the other systems in that it is primarily intended for download to the users' machine. It is designed to integrate any form of experimental data in a common format and is available for Sun, Solaris, DEC (OSF), and SGI (IRIX) machines; there is also a Mac version (MACACE). It stores gene-expression data as a part of a much wider range of information about *C. elegans*, in particular genetic- and physical-mapping data (clones and contigs), and the complete DNA sequence. The main interface is for UNIX computers and uses an X-windows-based, mouse-driven click-and-point navigation method. Users add their own data to the downloaded databases and for the public database, data is submitted to be entered by the database curator. The database accepts both text and original image data via e-mail or ftp. Text data are submitted as ASCII files that are read into the database in a standard tree-form structure. Images are added to a picture library and can be called from the database and displayed in a separate viewer (e.g. xv on UNIX versions). The system provides query and table-making functions, bibliography searches, and general search engines. The display capabilities include configurable genetic-map displays, physical-map displays and sequence-feature displays for DNA. Gene-expression data can be searched by text string or accessed through searches on the other types of data, including individual cells, cell groups, sequences, loci, clones and bibliographic information. This database is fully operational. However, most of the gene-expression data comes from just two laboratories and is not comprehensive. Submission by other workers is being encouraged.

An example of a freely accessible gene expression database, based on ACeDB, is given in Table 14.1 (second WWW address). The first page provides a set of cosmids, each is a link to a page describing which genes exhibit expression patterns and for which no expression has been observed. In each case, the pattern is described in text and in some cases pictures of the original data are available.

## 14.6.2    *Drosophila*

Three databases exist, or are being developed, to store gene-expression data relating to *Drosophila* development. The database of Hartenstein *et al.* (1995) aims to spatially map gene expression and other data to digital models of the *Drosophila* embryo. Each model is being reconstructed from differential interference contrast images of optical section planes taken at 2-μm intervals through embryos. All the major organs will be outlined on these images and users will be able to map data of labelled cells, gene expression, etc. onto these images of sections and thereby be visualized in the context of the three-dimensional model. The system is not yet operational as a database, however, two *Drosophila* gene expression databases that are currently accessible are described below.

## 14.6.2.1    *Flyview*

Flyview (Janning, 1997) is a database that holds gene-expression data for enhancer-trap lines and cloned genes and is accessible via the WWW. This data is text, images of original data and bibliographic data. Access to the data is provided by search of the pattern elements (textual terms), search of the stocks in terms of stock number, allele, genotype, chromosome detail, viability, developmental stage and expression patterns, and finally access through an index of the entire stocks from different sites. The text descriptions use as far as possible, the controlled vocabulary of Flybase (see Table 14.1) which is also used for the text search of the database. Individual researchers can submit their own data, published or unpublished, either via the WWW or by e-mail or ftp. The images can be submitted digitally or as slides or photographic prints and even as original data (microscope slides). Most of the database records are linked to Flybase and/or to the Encyclopaedia of *Drosophila*. The database is operational and the enhancer-trap lines are available to the community.

## 14.6.2.2   Flybrain

Flybrain is an online atlas and database of the *Drosophila* nervous system (Heisenberg & Kaiser, 1995) based around peer-reviewed descriptions of the central and peripheral nervous system of *D. melanogaster*. The atlas includes a variety of linked representations: schematic diagrams, silver-stained sections of the embryonic and adult brains, autofluorescence sections of the adult brain and a number of antibody and golgi preparations. Gene-expression patterns of specific gene products will be entered with reference to the atlas that allows patterns from different experiments to be directly compared. The data holds patterns from enhancer-trap lines as well as specific genes and mutants. Most of the data will relate to *D. melanogaster*, but information from other *Drosophila* species, and other Diptera, may be included for comparison. It is planned to link the database to Flybase. This database is operational, but has few gene-expression data.

## 14.6.3   Zebrafish

The Zebrafish Database (Westerfield *et al.*, 1997) currently provides a comprehensive view of developmental and genetic data to which it is planned to add gene-expression patterns. The database has an atlas for staging embryos, an anatomical atlas of the adult (including a standard anatomical nomenclature), information on mutants (including images), genetic information (gene names, cDNA- and genomic-sequence, markers, etc.), data on probes and antibodies, a bibliography and lists of people working on zebrafish. It is planned to store gene-expression data both as textual descriptions (using a controlled vocabulary) and image form. Unpublished as well as published data will be accepted, although unpublished information will be marked so that users can assess the confidence level of each entry. Data will be entered by 'authorized users' and by the database team. The database will be accessible through a graphical interface to the WWW. 'Guest' users will be able to run common types of searches directly from the WWW interface and more complex queries will be possible via the database team. Spatial comparison and search will be based on graphical representations of the embryo for which voxel models are being created. This database is not yet operational. It is planned, eventually, to link this database to those for other organisms, including the MGEIR.

## 14.6.4   *Xenopus*

The *Xenopus* Molecular Marker Resource (XMMR) provides a comprehensive set of links to all aspects of *Xenopus* development as well as the *Xenopus* research community. It lists genes and markers and documents their expression in the *Xenopus* embryo and links this information to other genetic and developmental data. The database entries are a set of web pages accessible either by search of the text content or by browsing the index. The index provides a straightforward list arranged in terms of tissue or staining pattern and is probably the easiest route to finding the entries of interest. Clicking on the gene name in the list brings up the associated entry. Each submission is generated as a web page (help available from Dr Peter Vize who runs the database). It includes descriptions in free text, images of whole mounts and sections, and bibliographic references linked to Medline. Images can be submitted electronically or as hard-copy, including photographs and projection slides. In most cases, the descriptions include links to sequence databases and addresses for available probes, antibodies, etc. The XMMR home page has links to a wide range of information on *Xenopus*, including standard stages, and lists of libraries and expression plasmids.

## 14.6.5   Mouse

There are a number of gene-expression databases that include data from mouse tissue. The Mouse Gene Expression Information Resource (MGEIR) is a composite database comprising several different modules that document gene expression in the mouse embryo. Parts of the database are now available via the WWW. At present, the database submission interfaces are only available to the database curators; however, it is intended that the integrated MGEIR will be accessible via platform-independent, textual and graphical interfaces. This database will be fully integrated with the Mouse Genome Informatics (MGI) database (formerly MGD) and the 3-D Atlas of mouse development (Baldock *et al.*, 1992; Ringwald *et al.*, 1994; Davidson *et al.*, 1999). The MGEIR comprises: the MGI Gene Expression Database (formerly GXD), the Mouse Graphical Gene Expression Database (GGED), the Edinburgh Mouse Atlas that includes the Mouse Embryo Anatomical

Database. Data will include patterns from *in situ* hybridization, histochemistry, immunohistochemistry and information from assays on homogenized tissues. Gene expression is documented as text using the controlled vocabulary of the Mouse Embryo Anatomy Database, as images of original data, and subsequently, as graphical data spatially mapped to the embryo models of the 3-D Mouse Atlas. Retrospective data will be entered by the database team. New data will be submitted by individual researchers and checked by the database team. Published and unpublished data will be marked accordingly and it is planned to offer, as a further alternative, a peer review of submissions in conjunction with one or more established journals so that appropriate entries can achieve full publication status. All the data may be searched by text-string and spatial searches. Parts of MGEIR are now operational.

The individual parts of the MGEIR are described below.

### 14.6.5.1   The Mouse Embryo Anatomy Database

To ensure consistency the anatomical descriptions used in the databases are based on a database of morphologically recognizable anatomical structures. These are organized in a spatial hierarchy for each of the 26 developmental stages defined by Theiler (1989). This database can be browsed as a set of text pages using the search facilities of the web browser or by using a Java interface. The Java interface is an early prototype and will work correctly within Internet Explorer but not Netscape. The data includes a thesaurus of alternative names, notes and references on contentious aspects of the nomenclature, information on tissue type/architecture. The database also holds, but does not yet make accessible, information in the derivative and progenitor tissues for each component (where reliable information exists) to enable the user to trace the origins and future development of each component. This 'anatomical' lineage may be superseded by true cell lineage data entered in the GGED. Comments on nomenclature and suggested additions can be addressed to Jonathan Bard (j.bard@ed.ac.uk).

### 14.6.5.2   The Edinburgh Mouse Atlas

To allow the mapping of gene-expression data into a common spatial framework, three-dimensional models of mouse embryos are being constructed from serial sections of the histology for successive stages of development. The model embryos, together with utility software, will be available as a set of CD-ROMs and via the WWW through a Java interface. The models will represent stages from fertilization to Stage 22 and will eventually include at least some older stages, including Stage 25. These grey-level, voxel models are generally from the same embryos as illustrated in *The Atlas of Mouse Development*

(Kaufman, 1994). The major named anatomical components in each model embryo will be delineated in each reconstruction and linked to the anatomical nomenclature. These anatomical regions or domains within the three-dimensional reconstructions provide the relationship between text or purely spatial descriptions of the gene-expression patterns. Software from the Edinburgh group will allow the reconstructed embryo to be resectioned digitally in any orientation in order to show 'photographic' views of tissue structure and to allow each delineated named tissue to be tracked through the embryo. Output of these tissues into standard three-dimensional formats will allow users to view the atlas with a number of visualization packages such as AVS (Advanced Visual Systems. Inc. http://www.avs.com/) or the Visualisation Toolkit (http://www.kitware.com/vtk.html), these allow a wide range of manipulations, including stereo three-dimensional visualization. The current status of the availability of the CD-ROM and associated software is provided in the Edinburgh Mouse Atlas web pages.

## 14.6.5.3  MGI Gene-expression Database

The gene-expression entries to MGI are now operational and can be searched under as gene expression data or as cDNA clone and expressed sequence tag data. The database stores and integrates data from assays of gene expression *in situ* and in homogenized tissue (including RNA *in situ* hybridization, immunohistochemistry, northern and western blots, RT-PCR, RNA-ase protection, cDNA arrays, etc.). The database is fully integrated with the Mouse Genome Database (now MGI) to foster a close link to genetic and phenotypic data of mouse strains and mutants. Results from a search provide a text description of the expression pattern with additional information available via links to other data in the database or to bibliographic references. Data are being acquired by annotation from the literature by database editors, and in the future by direct transfer of expression data from laboratories. Expression patterns are described in the controlled vocabulary of the Mouse Embryo Anatomy Database and, for *in situ* studies, digital images of original expression data. The submission interface is currently in prototype form and therefore not yet available but will feature validation tools to facilitate standardized data annotation and cross-referencing. The database supports complex textual queries, such as combinatorial queries, and searches on data additional to gene expression. The images of original data will be indexed via terms from the anatomy database to provide direct links between the expression domains observed and the standardized anatomical descriptions and to make the images accessible to global text queries. The MGI gene-expression database editors currently identify all newly published research articles documenting data on endogenous gene expression during mouse development. In a first step towards annotation of the data in the MGI, these articles are indexed with regard to authors, journal, genes and embryonic stages analysed, and expression assays used. The index is updated daily and is also available via the WWW (Table 14.1, MGEIR, Jackson site).

### 14.6.5.4   *Graphical Gene Expression Database (GGED)*

This database is not yet available but will store graphical gene-expression data using the three-dimensional images of the Mouse Atlas as a spatio-temporal framework. Data will be submitted by individual users via a WWW, Java-based electronic interface employing custom-designed software to map the gene-expression domains to the model embryo reconstructions from the Atlas. Entries submitted will be checked by the database team. Custom software will allow users to enter data by a combination of semi-automatic and manual graphical methods and text. Additional software will allow users to analyse their data locally, for example, to identify regions of expression by name, to view expression domains in three-dimensional relation to the structure of the embryo and to share data with collaborators. Data entered locally will be transmitted via the Internet for submission to the public database. Similarly, graphical queries will be made locally and sent to the remote database; the results will then be returned to be analysed in the context of the Mouse Atlas. The GGED database will be inter-operable with the MGI gene-expression database so that the additional data available from MGI can be linked with GGED submissions. This interoperability will enable users to access a wide range of information and to combine text and image methods to enter data and query the database.

## 14.6.6   Human

No databases of gene-expression in human are available but at least two databases are planned for *in situ* gene-expression information from human embryos. The database being developed at the Department of Human Genetics, University of Newcastle-upon-Tyne, UK will be publicly accessible via the WWW and will include text and image data. Part of this database is being built in collaboration with the MGEIR Edinburgh group and it is intended that this database should be interoperable with the MGEIR. To ensure consistency between the systems, the human anatomical nomenclature system (see Table 14.1) that is being built for this database at the Department of Anatomy, University of Edinburgh, UK, will be compatible with the Mouse Embryo Anatomy Database. The second human gene-expression database also relates to embryos and is based at the Institute of Child Health, London, UK, and is associated with a graphical atlas of human embryonic development being built at the University of St Andrews, UK. This database is not publicly available but a public database is planned which will hold images of gene expression patterns *in situ* as well as images and text descriptions documenting gene expression results from homogenized material. The textual aspects of this database will, as far as possible, be made compatible with the Mouse and Zebrafish Database through the use of the standard human embryo anatomical nomenclature being developed in association with the MGEIR.

## 14.6.7    Special systems

Two gene-expression databases have been developed that relate to the development of specific organs – the kidney and the tooth. These collate information from a number of species and, because most of the data is submitted by the user community, they are relatively comprehensive.

### 14.6.7.1    The Kidney Development Database

Gene-expression information relating to the developing kidney is organized in tabular form. It is one of a set of similar databases that are being constructed for other organs that develop by arborization of an epithelium, for example salivary gland, lung, and mammary gland (see the Organogenesis Database, Table 14.1). The Kidney Development Database and the Organogenesis Database can be accessed on the WWW. Submissions are sent in via email and entered by the database curator. The Kidney Development Database contains more than 90% of currently available information on gene expression, mainly from mouse, chick and *Xenopus*, but data from zebrafish and other animals are included. The database contains a series of tables corresponding to data relating to specific sets of genes (e.g. growth factors) that are arranged so that each row provides a simple description of the gene expression pattern in terms of the tissue type. Some original images are included as well as bibliographic references and text annotations. This resource also holds data on mutant phenotypes and the effects of pharmacological treatments on organ development in culture. Queries are supported by a number of indexes that can be searched using the page search facilities of the WWW browser (usually under the Edit menu of the browser), more sophisticated search methods are planned.

### 14.6.7.2    The Tooth Database

The Tooth Database stores gene expression in text form mainly from mouse, rat and human and can be searched by gene type, assay type, stage and region of expression. Data includes same information on probes and antibodies, distinguishes different types of assays and bibliographic data. The gene-expression patterns are described in terms of a controlled vocabulary associated with the dental tissues at each stage of development. To ensure consistency, the anatomical terms are defined by diagrams at each stage, these also provide a means of search by clicking on the diagrammatic representation. The entries are in text form and can be queried by selecting from the list of possible views of the database. These views relate to gene type (e.g. signalling molecules), assay type, anatomical component, developmental stage, species and finally tooth type. From the tables the user can click on a particular gene to obtain a set of standard diagrams illustrating the expression in different tissues (including expression in parts of tissues). Submissions are accepted via email (bibliographic reference) or by filling in a submission form which can then be posted or faxed. The data is then entered by the database curator.

## 14.6.8   Other databases

There are a number of other databases with gene-expression data for mouse and human. DbEST holds information on expressed sequence tags. TBASE stores published and unpublished data on transgenic animals and targeted mutations. The database includes phenotypic information, some of which is relevant to gene expression. The information is given in free-text format and entries are made by TBASE staff who request verification and completion by the authors. Searches are by text term and complex combinatorial searches that include aspects of the methods used, dates, etc. are possible by using a form-based query interface. The results of a search can be given as a table or a list.

## 14.7   CONCLUSION

Most gene expression databases are still under development and it will be some time before complex query facilities involving the spatial and temporal patterning will be possible. Many of the systems are based on an underlying spatial model (often voxel models) so that gene-expression data will eventually be mapped into a common framework. When this is possible, the current databases that are used mainly to make generally accessible, and to store, collate, and scan information, will be extended to more sophisticated possibilities. As more of genes are identified and their expression patterns characterized and entered, it will become common to use gene-expression databases to frame testable hypotheses regarding genetic interactions in development. In this phase, there is likely to be an increasing emphasis towards relating the data to our knowledge of other measures of gene function. This knowledge will include genetic and biochemical evidence about interactions between gene products, and evidence on the distribution of cellular and morphogenetic processes within the embryo.

In terms of bioinformatics resources, the gene-expression databases are relatively simple. It will be important that they are fully integrated so that queries can be posed that span multiple databases, not just of gene expression but sequence, mapping, protein structure, function and mutation/dysmorphology databases. This interoperability requirement represents a challenge to the bioinformatics community that must be met if the full potential of the information locked into these databases is to be realized.

## REFERENCES

Ahrens E.T., Readhead C., Laidlaw D.H., Brosnan C., Jacobs R.E. & Fraser S.E. (1997) *J. Neurochemistry* **69**, s124.

Baldock R.A., Bard J.B.L., Kaufman M.H. & Davidson D. (1992) *BioEssays* **14**, 501–502.

Davidson D. & Baldock R.A. (1999) In *Molecular Genetics of Early Human Development*, Strachan T., Lindsay S. & Wilson D. (eds). BIOS Scientific Publishers Ltd, Oxford.

Davidson D., Baldock R.A., Bard J.B.L., Kaufman M.H., Richardson J.E., Eppig J. & Ringwald M. (1999) In In Situ *Hybridization. A Practical Approach*. Wilkinson D. (ed.). IRL Press, Oxford.

Hartenstein V., Lee A. & Toga A.W. (1995) *Trends Genet.* **11**, 51.

Heisenberg M. & Kaiser K. (1995) *Trends Neurosci.* **18**, 481.

Janning W. (1997) *Sem. Cell Dev. Biol.* **8**, 469–475.

Kaufman M.H. (1994) *The Atlas of Mouse Development*, 2nd Edn. Academic Press, London.

Martinelli S.D., Brown C.G. & Durbin R. (1997) *Sem. Cell Dev. Biol.* **8**, 459–467.

Object Management Group (1995) *The Common Object Request Broker Architecture and Specification*. OMG Publications, Framinghan.

Ringwald M., Baldock R., Bard J., Kaufman M., Eppig J.T., Richardson J.E., Nadeau J.H. & Davidson D. (1994) *Science* **265**, 2033–2034.

Theiler K. (1989). *The House Mouse: Atlas of Embryonic Development* (second printing). Springer-Verlag, New York.

Westerfield M., Doerry E., Kirkpatrick A.E., Driever W. & Douglas S.A. (1997) *Sem. Cell Dev. Biol.* **8**, 477–488.

# 15 Using the EcoCyc Database

*Peter D. Karp*

Pangea Systems, Inc., Menlo Park, CA, USA

## 15.1 INTRODUCTION

The EcoCyc database describes the *Escherichia coli* genome, the metabolic pathways of *E. coli*, and other aspects of *E. coli* cell function. This chapter provides an overview of the information contained in EcoCyc, and of how to use the World Wide Web (WWW) version of EcoCyc to access this information (an X-windows version of EcoCyc is also available).

As an up-to-date reference source on the annotation of the *E. coli* genome, EcoCyc will be useful to researchers who are annotating other microbial genomes. Although EcoCyc provides functional annotations for all *E. coli* genes, the annotations of genes that encode enzymes, transport proteins, tRNAs, and two-component signal-transduction proteins are particularly detailed. EcoCyc encodes information about all known enzymes of *E. coli* small-molecule metabolism, and about the metabolic pathways that these enzymes catalyse. The biochemical information within EcoCyc can be used to predict the metabolism of other organisms from their genomes (Karp *et al.*, 1996), and serves as a reference source for biochemists.

Many biological databases, such as Genbank and Swiss-Prot, contain a single type of biological data: Genbank contains nucleotide sequences, Swiss-Prot contains protein sequences. In contrast, EcoCyc describes many different types of biological entities. Each biological entity described within EcoCyc is represented as a distinct frame (object) in the frame knowledge representation system that underlies EcoCyc (Karp & Paley, 1996). Table 15.1 shows the current number of objects in the current version (4.1) of EcoCyc.

The remainder of this chapter describes how to query EcoCyc to retrieve information about *E. coli*, and how to interpret the information pages that EcoCyc generates. For more information about EcoCyc, see Karp & Paley, (1996); Karp *et al.*, (1998).

*Genetics Databases*
ISBN 0-12-101625-0

**Table 15.1**   The number of objects in
several EcoCyc classes.

| | |
|---|---:|
| Reactions | 843 |
| Enzymes | 665 |
| Two-component signal transducers | 45 |
| Pathways | 141 |
| Genes | 4682 |
| tRNAs | 79 |
| Compounds | 1304 |

## 15.2   GENES

The 4682 genes in EcoCyc reflect both those genes identified by the Blattner laboratory in the full genomic sequence of *E. coli* (Blattner *et al.*, 1997), and additional *E. coli* genes reported in the literature. The exact position of many of these latter genes on the *E. coli* chromosome is not known because many were not sequenced by the authors of the publication: for only 4393 of the 4682 genes is the exact position of the gene on the chromosome recorded in EcoCyc. Most, if not all, of the remaining 289 genes are the same gene as one of the mapped genes, but biologists have not yet determined these correspondences. Over time, the EcoCyc project will merge together entries for corresponding genes as these identifications are reported in the literature. The EcoCyc project will also continue to update the functional identifications of *E. coli* genes as reported in the literature, and as determined by ongoing sequence analysis by the EcoCyc project.

The EcoCyc home page at URL http://ecocyc.PangeaSystems.com/ecocyc/ provides background information on the project, such as links to publications and documentation. Users can query EcoCyc genes through the WWW in several ways. Most queries begin at the EcoCyc Query Page at http://ecocyc.PangeaSystems.com:1555/server.html. The different sections of this page provide different ways to query the EcoCyc data.

The section entitled 'Query by name or EC number' allows the user to query a gene by name. The *fumA* gene, for example, can be queried as follows. First select the type of object to query. The default type is 'Enzyme'; use the mouse to change this selector button to 'Gene'. Then type 'fumB' in the text box, and click Submit. EcoCyc will return the page of information about this gene shown in Fig. 15.1.

The page contains the following information:

- *ID*: the internal identifier assigned to this gene within EcoCyc.

# *E. coli* Gene: *fumB*

Show Sequence     Show Reverse Complement Sequence     Query Genbank

ID: EG10357

Synonyms: b4122

Superclasses: TCA cycle

Map position (centisomes): 93.62

Map position (nucleotides): 4,344,904-> 4,343,258

Products: fumarase B monomer submit of fumarase B

Reactions catalyzed by enzymes:
fumarate + $H_2O$ = malate

Pathways involving enzymes: anaerobic respiration, fermentation, TCA cycle, aerobic respiration

Gene-Reaction Schematic:

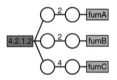

Unification Links: CGSC.18289

History: 10/20/97 Gene b4122 from blattner lab Genbank (v. M52) entry merged into EcoCyc gene EG10357; confirmed by SwissProt match.

Operon:

**Figure 15.1** The EcoCyc WWW page for the gene *fumB*.

- *Synonyms*: one or more synonyms for the gene name are listed, including the 'b-number' identifier assigned by the Blattner laboratory. (When querying by name the user can use any synonym, including the b-number.)
- *Superclasses*: this field lists the one or more classes to which the gene has been assigned in the Riley gene classification (Riley, 1993).

Clicking on the class name will return a page for that class, listing all genes within the class.

■ *Map position*: the map position of the gene within the *E. coli* chromosome. The first number is the centisome position (1 centisome unit equals one-hundredth of the chromosome length). The second two numbers give the starting and ending nucleotide positions of the gene (coding region). Click on the map position to see an EcoCyc page containing a view of the chromosomal region containing that gene.

■ *Transcription direction*: the direction of transcription, given as + (clockwise) or − (counter-clockwise).

■ *Products*: the name of the gene product, and the name(s) of any protein complex(es) that the product is part of. The product name is hot linked, meaning that clicking on the name of the gene product will return an EcoCyc page describing the product.

■ *Reactions catalysed by enzymes*: if the gene product is an enzyme, the one or more reactions catalysed by that enzyme are listed. The reaction is hot linked.

■ *Pathways involving enzymes*: if the gene product is an enzyme, the metabolic pathway(s) containing that enzyme are listed. Each pathway is hot linked.

■ *Gene-reaction schematic*: this diagram is a schematic depiction of the relationships between a gene, its protein products and reactions catalysed by those proteins. This diagram shows that *E. coli* contains three isozymes for fumarase. Three different genes (*fumA*, *fumB*, *fumC*) code for gene products (the circles they are connected to) that form protein complexes, shown in the second set of circles. The *fumA* and *fumB* complexes are homodimers; *fumC* forms a homotetramer. These protein complexes all catalyse the same reaction, whose EC number is 4.2.1.2. The *fumB* box is highlighted when drawn within the page for that gene. All of the elements of the schematic diagram are hot linked.

■ *Unification links*: this field contains hot links to other databases that contain information about this gene; in this case the link is to the Coli Genetic Stock Center (Berlyn & Letovsky, 1992).

■ *History*: this field contains comments about past changes that have been made to this EcoCyc frame. This particular history entry records the fact that when information from the full Blattner laboratory sequence of *E. coli* was loaded into EcoCyc, the EcoCyc project determined that gene b4122 identified by the Blattner laboratory was the same as the EcoCyc gene whose identifier was EG10357. Therefore, the data from the Blattner laboratory Genbank entry for this gene was merged into that EcoCyc gene frame. The gene correspondence was confirmed because the Swiss-Prot identifier attached to the EcoCyc gene and to the Blattner laboratory gene were the same.

- *Operon*: the operon organization of many *E. coli* genes will be shown in future EcoCyc versions.

At the top of the gene page are three buttons. Clicking on these buttons will: (1) display the nucleotide sequence of this gene from the Blattner laboratory Genbank record, (2) display the reverse complement of that sequence, or (3) will query Genbank via Entrez for all Genbank entries containing a gene called '*fumB*' in the species *E. coli*.

The bottom of the EcoCyc Query Page provides another way to query genes, under the heading 'Browse classification hierarchy'. To perform a query, first select the type of hierarchy to browse (in this case, the Genes hierarchy), then click on 'Submit'. The resulting page will list the categories of Riley's classification system (Riley, 1993), plus two extra categories: ORFs (genes whose function is unknown), and Unclassified (genes whose function is known but that have not yet been classified into one of Riley's categories). Clicking on any class name will return a page for that class that lists the superclasses of that class, the subclasses (if any) and the instances of the class (the genes assigned to that class); all of these items are hot linked.

## 15.3   PROTEINS

This section describes how to query enzymes and two-component signal-transduction proteins in EcoCyc. Although these two types of proteins are queried in the same way, their EcoCyc pages are interpreted differently.

### 15.3.1   Enzymes

Like genes, proteins in EcoCyc can be queried by name, in the 'Query by name or EC number' section of the query page. EcoCyc name queries perform substring searches as well as searching for exact matches. Therefore, if you select a Protein query, and enter 'pyruvate', EcoCyc will find all proteins whose common name (primary name) or whose name synonyms contain the string 'pyruvate'. One possible point of confusion is that although EcoCyc searches both the common name and the synonyms, the names that it prints in the list of matches are the common names, so it is possible that some of the listed matches will not actually contain 'pyruvate' because the match occurred in a synonym rather than in the common name.

All of the listed matches are hot links to EcoCyc enzyme pages, such as that shown in Figure 15.2.

The enzyme page is broken down into several sections: one section that

# E. coli Enzyme: phosphoenolpyruvate carboxykinase (ATP)

Superclasses: polypeptides

Gene: pckA

Molecular weight (kdaltons, from nucleotide sequence): 51.31 [1]

pI: 8.35

Neidhardt spot number: E056.0

Unification Links: Entrez.P22259, SWISS-PROT.P22259

Gene-Reaction Schematic:

| 4.1.1.49 |●— pckA |

---

**Enzymatic reaction of: phosphoenolpyruvate carboxykinase (ATP)**

Synonyms: phosphopyruvate carboxylase (ATP), phosphoenolpyruvate carboxylase, phosphoendpyruvate carboxykinase, ATP:oxaloacetate carboxy-lyase (transphosphorylating)

oxaloacetic acid + ATP <=> $CO_2$ + phosphoenolpyruvate + ADP

Reaction direction: REVERSIBLE

In pathways: gluconeogenesis

Comment: Phosphoenolpyruvate (PEP) carboxykinase is unusual in that it appears to be a monomeric enzyme with an allosteric site on its surface. [2]

Cofactors: $Mg^{+2}$

Alternative cofactors for $Mg^{+2}$:$Ca^{+2}$, $Mn^{+2}$

Activators (allosteric): $Ca^{+2}$ [2]

Inhibitors (allosteric): NADH

Inhibitors (mechanism undefined) [3]: ATP, phosphoenolpyruvate

Primary physiological regulators of enzyme activity: NADH

---

**Figure 15.2**   The EcoCyc WWW page for the enzyme phosphoenolpyruvate carboxykinase.

describes physical properties of the enzyme, followed by one or more sections that describe reactions catalysed by the enzyme, followed by one or more sections describing the subunits of the enzyme (if any). The enzyme page contains many of the same fields as the gene page (such as a gene-reaction schematic). The section describing the physical properties of the enzymes has the following additional fields:

- *Gene*: the gene that encodes this enzyme (this enzyme consists of a single gene product)
- *Molecular weight and pI* for this protein
- *Niedhardt spot number*: an identifier for this protein in the Eco2DBase DB (Van Bogelen *et al.*, 1992)

The section describing the reaction catalysed by the enzyme has several fields:

- *Name and synonyms*: an enzyme that catalyses multiple reactions will have different names for each reaction that it catalyses since enzyme names are derived from enzyme activities. The section for each reaction encodes those different names, as well as other information that is specific to the pairing of that enzyme and that reaction.
- *Reaction equation*: the reaction catalysed by the enzyme is listed, along with alternative substrates that the enzyme will accept, when known.
- *Reaction direction*: this reaction can be specified as reversible, meaning that it occurs in both directions in physiological settings; or irreversible, meaning it can occur only in the forward direction; or physiologically unidirectional, meaning that only the forward direction is known to be physiologically important.
- *In pathways*: the one or more pathways that the reaction participates in.
- *Cofactors*: one or more cofactors that the enzyme requires; alternative cofactors that can substitute for the primary cofactors are listed when known.
- *Activators and inhibitors*: Known activators and inhibitors of the enzyme are listed, broken down by mechanism, when known. Because some of these activators and inhibitors may have been determined *in vitro*, the database records which of these are the primary physiological regulators of enzyme activity.

The last section of the enzyme page lists the subunits of the enzyme, when known.

Several bracketed numbers are sprinkled throughout the enzyme page in Fig.

15.2, both embedded within comments, and attached to particular data values. These refer to both publications (some accessed through Medline and some stored within EcoCyc) and to comments (stored within EcoCyc). Citations attached to data values refer to the publication from which those data were obtained.

Enzymes can also be queried through the EC number system. However, because EC numbers are properly associated with reactions, rather than enzymes (Karp & Riley, 1993), EC numbers must be queried under the Reactions type. For example, under 'Query by name or EC number', select 'Reaction' and then enter 1.1.99.14 to query the reaction catalysed by the enzyme glycolate dehydrogenase. That query will return an EcoCyc reaction page, which will list the substrates of the reaction (including chemical structures) and the enzyme(s) that catalyse the reaction (as hot links). Click on a compound name to retrieve the EcoCyc page for that compound, which includes chemical information about the compound, as well as cross references to all reactions and pathways that contain the compound.

Alternatively, the entire EC classification system can be examined under the 'Browse Classification Hierarchy' section of the query page. Select 'EC Hierarchy' then 'Submit'. The resulting list will show the full EC number hierarchy; clicking on any of the EC classes will list its superclasses, subclasses and direct instances. The page that lists the EC number hierarchy includes other classes outside the EC system, such as 'Unclassified reactions' (reactions that have not yet been assigned an EC number), and '2Comp-Reactions' (reactions of *E. coli* two-component regulatory proteins).

## 15.3.2    Other proteins

Two-component signal transducers, transporters and other proteins can be queried by name just as enzymes can. For example, under 'Query by name or EC number', enter the name NRI to query the NRI regulator. The resulting page contains a list of proteins whose name contains the string 'NRI'; click on NRI to see the page for the NRI protein, shown in Fig. 15.3. NRI is involved in regulation of enzymes for nitrogen metabolism. The NRI page describes the NRI protein and lists reactions that the protein occurs in as a substrate. EcoCyc describes signal-transduction events as reactions. Two reactions are listed for NRI: in the first NRI-P dephosphorylates autocatalytically; in the second, a phosphate group is transferred from NRII-P to NRI. Thus, EcoCyc protein pages encode information about proteins both as catalysts and as substrates. Click on 'NRI-PWY' to see a drawing of the signal-transduction pathway in which NRI is involved.

# E. coli Protein: NRI

Synonyms: transcription factor, nitrogen regulator protein I

Superclasses: nr-i

Component composition: NRI x 2

Comment: In the unphosphorylated state NRI is a dimer, phosphorylated it forms a tetramer. [1, 2]

In Reactions:

NRI-PWY:
$H_2O$ + NRI-P = phosphate + NRI,
NRII-P + NRI = NRII + NRI-P

Gene-Reaction Schematic:

**Subunit: NRI**

Synonyms: nitrogen regulator protein I, unmodified nitrogen regulator protein I

Gene: glnG

Molecular weight (kdaltons, from nucleotide sequence): 52.254 [3]

pI: 6.52

**Figure 15.3**  The EcoCyc WWW page for the NRI protein.

## 15.4 PATHWAYS

EcoCyc describes all known pathways of *E. coli* small-molecule metabolism. EcoCyc pathways can be queried in several ways. The EcoCyc query page allows pathways to be queried by name/substring, and provides access to a hierarchical system of pathway classification – the same types of queries available for genes. In addition, the query page provides a menu of all EcoCyc pathways; simply select a pathway from that menu and click Submit to see information about that pathway, such as the page shown in Fig. 15.4.

Pathway diagrams show a linked set of reactions. For each reaction the information shown includes the EC number, the enzyme(s) that catalyse the reaction, the gene(s) encoding the enzyme(s), and the substrates. Enzymes that are sensitive to activators or inhibitors are tagged with a circled plus or minus

# E. coli Pathway: rhamnose catabolism

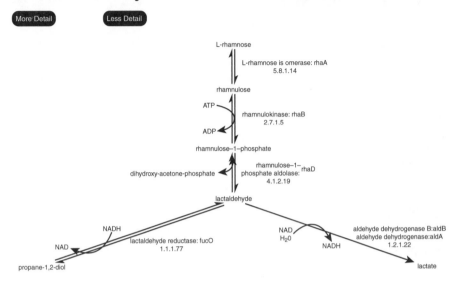

Synonyms: rhamcat

Superclasses: Carbon compounds

Net reaction equation: rhamnose + ATP = glycerone phosphate + lactaldehyde + ADP

Superpathways: fucose and rhamnose catabolism

Locations of Mapped Genes:

**Figure 15.4**   The EcoCyc WWW page for the rhamnose catabolism pathway.

sign. All of these elements of the pathway drawing are hot links, e.g., clicking on a substrate name will display the EcoCyc compound page for that metabolite. Other information present in pathway pages includes:

- *Net reaction equation*: the net chemical transformation accomplished by the pathway.
- *Superpathways*: the one or more superpathways of which this pathway is a part. A superpathway is simply a linked set of smaller pathways that shows how those smaller pathways are interconnected.

- *Locations of mapped genes*: this diagram shows the location on the *E. coli* chromosome of the genes whose products catalyse steps in the pathway. Moving the mouse pointer over the mark for a gene will display the gene name.

At the top of the pathway page are buttons labelled 'More Detail' and 'Less Detail'. Clicking on those buttons will produce different views of the same pathway. The view with more detail will show the chemical structures of pathway substrates. Typically, four levels of detail are available for a given pathway.

A diagram showing the full metabolic map of *E. coli* is also accessible from the query page by clicking on the phrase "Metabolic Overview diagram." The curves in this diagram are substrates and the lines are enzyme-catalysed reactions. Move the mouse over one of the substrates to identify it. The X-windows version of EcoCyc provides additional ways of interrogating this diagram, such as finding a reaction or pathway, or highlighting all reaction steps that are inhibited by a given compound.

## 15.5 SUMMARY

The types of biological entities described within EcoCyc include: chromosomes, genes, gene products (with a current focus on enzymes, tRNAs and two-component signal-transducers), reactions, chemical compounds and pathways. EcoCyc provides query and visualization tools for each of these entities.

Information within EcoCyc can be found directly or indirectly. For the direct approach, use the section of the query page corresponding to the type of query you wish to make: a name-based query or a query according to classification hierarchy. Next, select the type of object you wish to query, such as a gene, and then enter the name you wish to query. For the indirect approach, first use the direct approach to find an object (such as a chemical compound) related to the object you wish to find (such as an enzyme catalysing reactions involving that compound) and then use the hot links to navigate to the object of primary interest.

## ACKNOWLEDGEMENTS

This work was supported by grant 1-R01-RR07861-01 from the National Center for Research Resources, and by Pangea Systems Inc. EcoCyc is joint work with Monica Riley of the Marine Biological Laboratory.

# REFERENCES

Berlyn M. & Letovsky, S. (1992) *Nucl. Acids Res.* 20, 6143–6151.

Blattner F., Plunkett III G., Bloch C., Perna N., Burland V., Riley M., Collado-Vides J., Glasner J., Rode C., Mayhew G., Gregor J., Davis N., Kirkpatrick H., Goeden M., Rose D., Mau B., and Shao Y. (1997) *Science* 277, 1453–1462.

Karp P. & Paley S. (1996) *J. Comput. Biol.* 3(1), 191–212.

Karp, P. & Riley, M. (1993). In *Proceedings of the First International Conference on Intelligent Systems for Molecular Biology*, Hunter L., Searls D. & Shavlik J. (eds). AAAI Press, Menlo Park, pp. 207–215.

Karp P., Ouzounis C. & Paley S. (1996) In *Proceedings of the Fourth International Conference on Intelligent Systems for Molecular Biology*, States D., Agarwal P., Gaasterland T., Hunter L. & Smith R. (eds). AAAI Press, Menlo Park, pp. 116–124.

Karp P., Riley M., Paley S., Pellegrini-Toole A. & Krummenacker M. (1998) *Nucl. Acids Res.* 26, 50–53.

Riley M. (1993) *Microbiol. Rev.* 57, 862–952.

Van Bogelen R., Sankar P., Clark R., Bogan J. & Neidhardt F. (1992) *Electrophoresis* 13, 1014–1054.

# Appendix: List of URLs in Text and Tables

| URL | Page no./Table no. |
|---|---|
| http://imgt.cnusc.fr:8104/ | Table 2.1 |
| http://transfac.gbf.de/TRANSFAC/ | Table 2.1 |
| http://condor.bcm.tmc.edu/oncogene.html | Table 2.1 |
| http://www.imb-jena.de/RNA.html | Table 2.1 |
| http://ndb-mirror-2.rutgers.edu/ | Table 2.1 |
| http://www.expasy.eh/prosite/ | Table 2.1 |
| http://www.biochem.ucl.ac.uk/bsm/dbbbrowser/PRINTS/PRINTS.html | Table 2.1 |
| http://www.sanger.ac.uk/Software/Pfam/ | Table 2.1 |
| http://protein.toulouse.inra.fr/prodom.html | Table 2.1 |
| http://base.icgeb.trieste.it/sbase/ | Table 2.1 |
| http://www.ncgr.org/gsdb/ | Table 2.1 |
| http://www.gcrdb.uthscsa.edu/ | Table 2.1 |
| http://rebase.net.com/ | Table 2.1 |
| http://immuno.bme.nwu.edu/ | Table 2.1 |
| http://www.rcsb.org/pdb | Table 2.1 |
| http://www.cbs.dtu.dk/databases/OGLYBASE/ | Table 2.1 |
| http://www.ncbi.nlm.nih.gov/dbEST/ | Table 2.1 |
| http://www.ncbi.nlm.nih.gov/dbSTS/ | Table 2.1 |
| http://www.ncbi.nlm.nih.gov/UniGene/ | Table 2.1 |
| http://www.mrc-spe.cam.ac.uk/imt-doc/public/INTRO.html | Table 2.1 |
| http://biomaster.uio.no/cpgisle.html | Table 2.1 |
| http://www.uwcm.ac.uk/uwcm/mg/hgmd0.html | Table 2.1 |
| http://www.mcs.anl.gov/home/gaasterl/genomes.html | Table 2.1 |
| http://www-fp.mcs.anl.gov/~gaasterland/genomes.html | Table 2.4 |
| http://golgi.harvard.edu/ | Table 2.4 |
| http://www.ebi.ac.uk/biocat/biocat.html | Table 2.4 |
| http://www.hgmp.mrc.ac.uk/genomeweb | Table 2.4 |
| http://www.biosupplynet.com/cfdocs/btk/btk.cfm | Table 2.4 |

*Chapter 3*

| | |
|---|---|
| http://www.informatics.jax.org | 39 |
| http://genome-www.stanford.edu/Arabidopsis/ | 39 |
| http://www3.ncbi.nlm.nih.gov/omim/ | 42 |
| http://www.cf.ac.uk/uwcm/mg/hgmd0.html | 42 |
| http://gdbwww.gdb.org/ | 43 |
| http://www.citi2.fr/GENATLAS/welcome.html | 43 |
| http://cedar.genetics.soton.ac.uk/public_html/ | 43 |
| http://www.angis.su.oz.au/Databases/BIRX/omia/ | 43 |
| http://www2.ebi.ac.uk/mutations/ | 44 |
| http://www.ebi.ac.uk/ebi-docs/swissprot_db/swisshome.html | 44 |
| http://globin.cse.psu.edu/ | 45 |
| http://www.genet.sickkids.on.ca/cftr/ | 46 |
| http://www.mcgill.cs/pahdb/ | 46 |
| http://www.umds.ac.uk/molgen/haemBdatabase.htm | 46 |
| http://perso.curie.fr/tsoussi/p53_database.html | 47 |
| http://www.iarc.fr/p53/Homepage.htm | 47 |
| http://sunsite.unc.edu/dnam/mainpage.html | 47 |
| http://www.mayo.edu/research/papers/P53%20Mutations | 47 |
| http://www.gen.emory.edu/mitomap.html | 47 |

| URL | Page no/Table no. |
|---|---|
| ftp://ftp.virginia.edu/pub/fasta/ | Table 8.1 |
| ftp://ncbi.nlm.nih.gov/pub/esr/dotter/ | Table 8.1 |
| ftp://ncbi.nlm.nih.gov/pub/tatusov/dust/ | Table 8.1 |
| ftp://ncbi.nlm.nih.gov/pub/pub/seg/ | Table 8.1 |
| ftp://ncbi.nlm.nih.gov/repository/repbase/censor/ | Table 8.1 |
| http://blast.wustl.edu/pub/xblast/ | Table 8.1 |
| ftp://ncbi.nlm.nih.gov/pub/neuwald/gibbs11_93/ | Table 8.1 |
| ftp://beagle.colorado.edu/pub/Consensus/ | Table 8.1 |
| http://www.cse.ucsc.edu/research/compbio/sam.html | Table 8.1 |
| ftp://beagle.colorado.edu/pub/ | Table 8.1 |
| ftp://ariane.gsf.de/pub/unix/matinspector | Table 8.1 |
| http://hmmer.wustl.edu/ | Table 8.1 |
| ftp://biosci.umn.edu/pub/proscan/ | Table 8.1 |
| ftp://beagle.colorado.edu/pub/PromFD.tar | Table 8.1 |
| http://www.genetics.wustl.edu/eddy/software/#rnabob | Table 8.1 |
| http://www.genetics.wustl.edu/eddy/tRNAscan-SE/ | Table 8.1 |
| http://www.snv.jussieu.fr/cgi-bin/wrap/viari/Palingol | Table 8.1 |
| http://www.scripps.edu/pub/goodsell/research/bend | Table 8.1 |
| ftp://sgjs1.weizmann.ac.il/pub/Curvature/ | Table 8.1 |
| ftp://expasy.hcuge.ch/pub/lalnview | Table 8.1 |
| ftp://ftp.isrec.isb-sib.ch/sib-isrec/SEView | Table 8.1 |
| http://www.charon.girinst.org/~server/repbase.html | Table 8.2 |
| ftp://ncbi.nlm.nih.gov/repository/repbase/REF/simple.ref | Table 8.2 |
| http://mbcr.bcm.tmc.edu/smallRNA/smallrna.html | Table 8.2 |
| http://pbil.univ-lyon1.fr/databases/hovergen.html | Table 8.2 |
| http://www.epd.unil.ch | Table 8.2 |
| http://transfac.gbf.de/TRANSFAC/ | Table 8.2 |
| http://www.mgs.bionet.nsc.ru/systems/TRRD | Table 8.2 |
| http://www.isbi.net/ | Table 8.2 |
| ftp://www.cifn.unam.mx/pub/software/mac/ | Table 8.2 |
| ftp://beagle.colorado.edu/pub/imd_1.1.tar.gz | Table 8.2 |
| ftp://transfac.gbf.de/pub/structure_library/ | Table 8.2 |
| ftp://ftp.ebi.ac.uk/pub/databases/nucleosomal_dna/ | Table 8.2 |
| http://c3.biomath.mssm.edu/trf.upload.form.html | Table 8.3 |
| http://bioweb.pasteur.fr/seqanal/interfaces/satellites.html | Table 8.3 |
| http://ftp.genome.washington.edu/cgi-bin/RepeatMasker | Table 8.3 |
| http://transfac.gbf.de/cgi_bin/patSearch/patsearch.pl | Table 8.3 |
| http://www.gsf.de/cgi_bin/matsearch.pl | Table 8.3 |
| http://agave.humgen.upenn.edu/utess/tess31 | Table 8.3 |
| http://bimas.dcrt.nih.gov:80/molbio/matrixs/ | Table 8.3 |
| http://transfac.gbf.de/dbsearch/funsitep/fsp.html | Table 8.3 |
| http://biosci.cbs.umn.edu/software/proscan/promoterscan.htm | Table 8.3 |
| http://irisbioc.bio.unipr.it/pol3scan.html | Table 8.3 |
| http://www.gsf.de/cgi-bin/fastm.pl | Table 8.3 |
| http://www2.icgeb.trieste.it/~dna/bend_it.html | Table 8.3 |
| http://www.genetics.wustl.edu/eddy/tRNAscan_SE/ | Table 8.3 |
| http://avalon.epm.ornl.gov/Grail-bin/EmptyGrailForm | Table 8.3 |
| http://www.itba.mi.cnr.it/tradat/ | Table 8.3 |
| http://menu.hgmp.mrc.ac.uk/Nix/ | Table 8.3 |

# Index